Product Maturity 1

Reliability of Multiphysical Systems Set

coordinated by
Abdelkhalak El Hami

Volume 12

Product Maturity 1

Theoretical Principles and Industrial Applications

Franck Bayle

iSTE

WILEY

Foreword by Laurent Denis

Human beings are plagued by major worries, such as fear of death and fear of illness. "How long will I live?" is a question that arises even in childhood. "Will I one day have to deal with a condition similar to my neighbor's?". We live in an age where disease, death, old age and disability are subjects to be avoided in polite conversation. "How are you?" is a standard greeting to which a different and darker reply than the traditional, "I'm very well, thank you, and you?" risks embarrassing or even annoying the other party. Avoiding the problems of others, for fear they may be contagious, gives us a sense of immortality on a daily basis.

This is a rather recent phenomenon, as many previous generations did not hide the elderly or sick, although the risk of accidents in everyday life was higher and so death was a more common occurrence. It was certainly a source of anxiety, but the Church was there to alleviate it. Today we hide this subject by paying attention to a society made up of young, healthy people whom we must emulate at all costs so as to be part of it. Since our days are more or less the same, we succumb to procrastination at the first opportunity and Seneca's *carpe diem* loses its wonderful charm to give way to flat Platonic reflection.

Surprisingly, a similar problem exists in industry: there is a willingness to forget that a product may be subject to failure during its lifetime, given it has been optimally designed for the required functions. Some simple principles of upstream reliability analysis, from the design phase onwards, are now well-established, but they thwart the deep-seated notion that proper design outweighs everything else. Two essential points are overlooked: when a technology naturally reaches maturity, only a technological breakthrough can mark a distinction between two products performing the same function, unless it can be demonstrated that product A will last longer and be safer than product B. Moreover, the uses of the same product can multiply according to its ability to adapt to multiple environments. A good understanding of these uses in the field makes it possible to improve robustness properly at the design stage, in order that it can withstand any mission profile assigned to it during operation; this is one way to increase competitiveness.

Many companies still see the reliability study of a system before it becomes operational as a mandatory step to be overcome, bypassing or minimizing it as soon as possible. In the design phase, a signed product FMECA will end up in a folder, its purpose merely to certify that the rules have been followed correctly. The objective of the test phase is to confirm that the device being tested meets the requirements of a standard, without taking the opportunity to validate that the mission profiles on the ground will not unpredictably damage the product. During production, process control cards are used to verify that tolerance limits are not exceeded, without establishing forecasting instances that could lead to accidental stops. Hence, only data in the form of returned products, found to be defective by the end user, are subjected to *a posteriori* analyses by customer support. This can incur various costs and may lead to product recall if a serious defect is found.

Fortunately, however, the reality tends to be a little less bleak than the situation described above, with the emergence and dissemination of best practices that are based on theories validated by various industry sectors. These are now adapting to the challenges that companies face: making increasingly complex products that are more adaptable and ever-faster, while maintaining quality standards and reducing costs. This no longer involves applying deterministic models in which a single value is assigned to an objective to be reached. Instead, it is about drawing up a range of possible solutions that allow the supplier or integrator to make sure that the worst case a product might be subjected to on the ground can still be controlled by statistical modeling. The best way to achieve this is through the combined use of theoretical and technical resources: an in-depth understanding of the possible technological problems and solutions given by the manufacturer allows the qualified reliability engineer to build the most suitable predictive models. Ideally, a single person would have these two complementary sets of skills.

Franck Bayle is a perfect example of this. Throughout the second part of his career as an electronics engineer, he relentlessly addressed challenges that no one had previously openly solved, and he developed algorithmic solutions based on cutting edge theories. He was nevertheless confronted with the ills that plague most large groups: habit and fear of change. When he proposed significant advances across the whole company, only his more informed colleagues considered these to be opportunities for improvement. Sometimes his work was considered useless by those whose feeling was: "Why consider risks when there are no problems on the ground?". This is reminiscent of: "Why would I get sick when I am fit and healthy?". We have to be forward thinkers to be able to act before any problem arises, and Franck Bayle is such a person. His book presents all the best practices he has managed to implement within his department, as well as all the advances that I have had the chance to see implemented, which he continues to improve.

This book is essential reading for any passionate reliability engineer, and it is a real pleasure and an honor to write this foreword to accompany it.

Laurent DENIS
STATXPERT
November 2021

Foreword by Serge Zaninotti

When Franck invited me to work with him on his second book on system maturity, I immediately accepted. My interest in the subject has grown largely as a result of the rich technical exchanges we have had over the last 15 years, and strengthened after reading his first book, published in 2019, on the reliability of maintained systems under aging mechanisms.

Franck would tell me of his progress in the field of reliability, his field of expertise, and I – having always wanted to maintain the link between quality and reliability – would try to establish a connection with the standards.

Indeed, thanks to those who trained me as a quality engineer, I have always known that quality assurance should never be dissociated from dependability. I therefore felt instantly motivated by the opportunity to contribute to disseminating the acquired knowledge by means of a book. The subject system maturity can be mastered both through experience and through training.

It is often the failures or non-quality observed during the development or operation of a system that indicate to us that our patterns of thinking lack dimension.

However, in order to find an appropriate response to prevent these unexpected and feared events, and to be able to control them in the best way possible when they do occur, it is important to master quality risk management techniques. Risk management begins with risk prevention, the focus of this book.

In order to understand the problem of system maturity as a whole, before addressing the actual techniques used, it is necessary to put it in context. This context is provided by the quality standards for the systems.

Having trained as a general engineer within the Department of Energy and Environmental Engineering (GEn) at INSA Lyon, I then gained experience as a quality specialist, and have been a dependability supervisor since 1989. Franck therefore asked me to present the standards environment and the links that tie it to maturity, which the reader will find in Chapter 2 of this book.

Serge ZANINOTTI
Thales
Quality Expert
November 2021

Acknowledgements

This book would certainly not have been possible without the contribution of certain persons. I therefore want to thank, first, my main supervisors throughout my career with Thales: Jean Riaillon, Laurent Portrait and Claude Sarno, who gave me the means to gain this experience.

For everything related to maturity, a special thank you goes to Serge Zaninotti, quality expert with Thales, and also the author of Chapter 2 on the notion of maturity and the "quality" aspects, and Serge Parbaud of Thales for his advice and always appropriate corrections. I would also want to extend my warmest thanks to Patrick Carton from Thales Global Service for the passionate technical exchanges we have had in recent years, his always apt remarks, his support and his listening.

Furthermore, I wish to thank Franck Davenel from DGA for our exchanges during PISTIS upstream study related to accelerated tests and burn-in, and to give my warmest thanks to Léo Gerville Réache for his valuable help.

Finally, I wish to thank my entire family, and particularly my wife, not only for bearing with me, but also for encouraging me while writing this book.

Introduction

Reliability, availability, safety and so on are now major qualities that a product must have, irrespective of the industrial application field (automobile, avionics, rail, etc.) of its use. A significant literature related to these fields can be readily accessed, and is generally grouped under the umbrella concept of "dependability".

During the whole lifecycle of a product, from specification to operation by the end user, a large number of actions are implemented in order for it to meet the specified requirements. Reliability is the quantitative basis for dependability activities, as poor reliability can lead to insufficient availability, for example, although it should be reached as soon as the products are in service.

The maturity of a product is therefore its capacity to reach the desired reliability level, from its launch into service until the end of its operation. Due to technical and economic challenges, it is very difficult to reach product maturity. Indeed, defects are very often generated during various phases of the lifecycle, reflected by failures that occur very early on in product operation (a manufacturing defect, for example), or during its operation (design flaw, integration flaw, etc.). This is particularly true for products whose service life is becoming longer (e.g. 30 years for components in the rail industry). It is important to note that this activity makes sense for maintained products, which are predominantly in industrial applications.

There is abundant information on maturity, but this applies mostly to process implementation within a company, and it is therefore often at the project management level. Detailed literature describing the main theories (worst-case analysis, derating analysis, etc.) and practical techniques (accelerated tests, burn-in, etc.) for building product maturity is actually scarce, and many manufacturers often use obsolete standards, which, at best, they modify according to their experience.

The main objective of this book is to fill this knowledge gap, which is often detrimental to many manufacturers.

1

Reliability Review

In this book, maturity is defined as the ability of a product to achieve the expected level of reliability from the moment it becomes operational for the end user. A review of what reliability means and a definition of the parameters on which it is based is therefore needed.

1.1. Failure rate

Reliability studies the occurrence of failures in time. These instances of failure are random; hence, they cannot be known in advance. This presents a challenge. To model them, we use the concept of random variable, which will be denoted by T throughout this book.

First, it is important to determine the various types of failures. There are three main categories, namely:

– "youth failures", which generally occur very early on in the lifecycle of a product. Youth failures are generally the result of manufacturing defects. Therefore, they concern only a small part of the population. They can be partially eradicated by specific tests, such as burn-in;

– "catastrophic failures", which are unexpected, sudden and independent of the time previously elapsed. These types of failures can therefore be observed at any point in the lifecycle of a product. They are generally the result of accidental overloads (heat, mechanical, electrical). They typically do not concern the entire product population and can be reduced by robustness tests, derating rules, etc.;

– "aging" failures, which are observed across all the products in operation. These failures are generally not observed during the lifecycle of a product, with the exception of specific components with a "limited service life" or premature aging, as a result of poor sizing, a batch of defective components, etc. They affect the entire population and therefore must be absolutely pushed beyond the duration of use of the product. Consequently, design rules (derating rules, worst-case analysis, thermal, mechanical, electrical simulation, etc.), and specific aging tests can be implemented.

We begin by addressing intrinsic reliability. Intrinsic reliability refers to the reliability of a component, a card or a product in the absence of any maintenance. In order to estimate this, and in particular to know the type of failure involved, the most widely used parameter is the (instantaneous) failure rate denoted by λ, which is defined by:

$$\lambda(t) = \lim_{\Delta t \to 0} \frac{P[t \leq T < t + dt / T > t]}{\Delta t} \qquad [1.1]$$

Let us briefly analyze this equation and the following conventions. The term P denotes the "probability" and the symbol "/" stands for "knowing that". The limit "lim" represents the instantaneous character of the failure rate. Therefore, equation [1.1] can be interpreted as follows:

Probability that the product will fail between "t and t+dt" knowing that it was operational (non-defective) at instant "t".

To facilitate understanding of the concept of failure rate, the analogy with a human being can be used (Gaudoin and Ledoux 2007). Let us try to estimate the probability that a human being dies between 100 and 101 years of age. This probability is low since the majority of human beings die before they reach 100 years old. Furthermore, let us estimate the probability that a human being dies between 100 and 101 years of age, knowing that they were alive at 100 years old. This probability is high, as human beings do not live long after reaching 100 years of age.

The three failure categories can thus be symbolically represented using the concept of failure rate using the famous bathtub curve, as illustrated in the following figure.

Figure 1.1. *Bathtub curve example*

The most commonly used mathematical object for modeling failure rate is the Weibull distribution. According to this hypothesis, the latter is defined by:

$$h(t) = \left(\frac{\beta}{\eta}\right) . \left(\frac{t}{\eta}\right)^{\beta-1} \qquad [1.2]$$

where η is a scale factor (generally time-dependent) and represents typical service life, characterized by the fact that the failure rate is ~ 63.2% (1 – exp(-1)), irrespective of the value taken by the parameter β and therefore of the type of failure.

This modeling is interesting for the following three reasons:

– the mathematical formulation is simple, as it involves a versatile power function (differentiable, integrable, etc.);

– depending on the parameter β, this function is decreasing ($\beta < 1$), constant ($\beta = 1$) or increasing ($\beta > 1$). In other terms, it can represent the three types of previously defined failures;

– the parameter β has a physical significance as it represents the aging dynamics of the observed failure mechanism. Indeed, as already noted, failure instants are characterized by randomness (components tested are assumed to be identical). This means that instead of having a single real value, if failures were purely deterministic, we see a constant dispersion of failure instants. In fact, the parameter β is the image of this dispersion, and the greater it is, the less dispersed the instants of failure are. Ultimately, if β was infinite, all the failure instants would be identical, which is obviously never the case in practice.

Figure 1.2. *Fall leaves illustrating aging. For a color version of this figure, see www.iste.co.uk/bayle/maturity1.zip*

This figure clearly shows that all of the components – in this case, the leaves – are subject to aging, yet not all of them fail at the same time (not all the leaves have fallen at the instant shown).

As an illustration, let us assume a Weibull distribution whose scale factor is $\eta = 1{,}000$ (this value is a purely conventional value and could be quite different without changing the conclusions obtained). Furthermore, let us assume that there are 30 components in a test and failure instants are generated for each of them in a purely virtual manner for two values of β (3 and 10).

The following figures are obtained, with time on the ordinate (horizontal) axis and the number of components on the abscissa (vertical) axis.

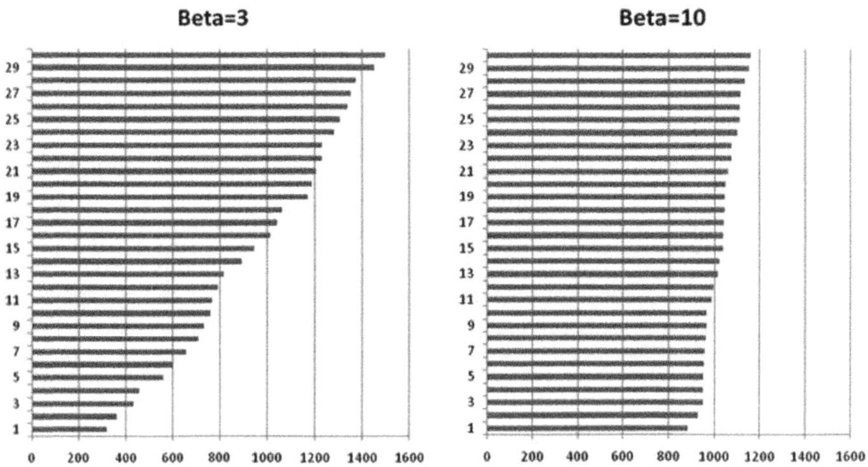

Figure 1.3. *Failure instants for $\beta = 3$ and $\beta = 10$*

It can be noted that failure instants are more dispersed for $\beta = 3$ (on the left) than for $\beta = 10$ (on the right). On the other hand, for $\beta = 1$, equation [1.2] is written as: $\lambda(t) = \frac{t}{\eta}$ or, more frequently, as:

$$\lambda(t) = \lambda \qquad [1.3]$$

This represents the exponential distribution law modeling catastrophic failures. The failure rate for this category of failures is constant, which means that failure instants do not depend on the elapsed time. This specificity of the exponential law is known as the "memoryless property" (it is the only continuous law with this property). Indeed, returning to the analogy with human beings, a catastrophic failure is, for example, a car accident occurring when a driver cuts off another driver. This "failure" does not depend on the distance traveled, but is due solely to the recklessness of another person. This is entirely different from an aging failure, for which the failure instant directly depends on the distance traveled, because this relates to driver fatigue.

It is important to note that the concept of maturity has no qualitative meaning for non-maintained products. Indeed, the objective of reliability is a probability of success; the mission is achieved by the survival function, which for a Weibull distribution is defined as:

$$R(t) = \exp\left(-\left(\frac{t}{\eta}\right)^{\beta}\right) \quad [1.4]$$

This survival function – and this is the case regardless of the law used – is a strictly decreasing function of time. Therefore, the concept of constant reliability is not applicable. For most non-maintained industrial applications, exponential distribution is preferred to Weibull distribution; this is because the reliability objective is a probability of achieving the mission, whose value is obviously high (generally such that $R \in [90\% ; 99\%]$.

In this case, we can return to an exponential distribution because, for these values of the survival function, it is conservative, with respect to a Weibull distribution, whose shape parameter is greater than 1. Indeed, from a mathematical perspective, the ratio of the two survival functions can be calculated as follows:

$$\zeta = \frac{\exp\left(-\left(\frac{Tm}{\eta}\right)\right)}{\exp\left(-\left(\left(\frac{Tm}{\eta}\right)\right)^{\beta}\right)} \quad [1.5]$$

with $\beta > 1$ and Tm = mission duration.

To obtain a sufficiently high probability of success in the mission (survival function) requires $Tm/\eta << 1$. Consequently, β being greater than 1, $\left(\frac{Tm}{\eta}\right)^{\beta} \ll 1$. Using an expansion up to the first order of the exponential function leads to:

$$\zeta \simeq \frac{1-\frac{Tm}{\eta}}{1-\left(\left(\frac{Tm}{\eta}\right)\right)^{\beta}} \quad [1.6]$$

Since Tm/η is greater than $\left(\frac{\mathrm{Tm}}{\eta}\right)^{\beta}$, the numerator is smaller than the denominator and therefore $\zeta < 1$. Hence, the exponential survival function is lower than that of Weibull, which proves that it is conservative.

Another, more physical way to view this result is to remember that the shape parameter β represents the dispersion of time until failure. The greater β is, the less dispersed the time until failure. Since the Weibull shape parameter is > 1, the corresponding failure instants are less dispersed around the scale parameter η.

Consequently, failures following an exponential distribution with an identical scale parameter occur earlier than those following the Weibull distribution. Therefore, the survival function of the exponential distribution at any instant "t" is weaker than that of the Weibull distribution, which proves the conservative character of this approach.

1.2. Temperature effect

Temperature is systematically involved in component failure mechanisms. The Arrhenius law is generally used in order to model its effect on the reliability of components. Based on an empirical research method, that is, studied through a number of experiments, the Arrhenius law is used to model the variation in the speed of certain chemical reactions under the influence of temperature. With respect to the previously described Weibull law, the following formulation is obtained:

$$\lambda(t,\theta) = \left(\frac{\beta}{\eta(\theta)}\right).\left(\frac{t}{\eta(\theta)}\right)^{\beta-1} \qquad [1.7]$$

with $\eta(\theta) = \mathrm{C}.\exp\left(\frac{\mathrm{Ea}}{\mathrm{Kb.T}}\right)$

Ea is the activation energy.

Kb is the Boltzmann constant.

1.3. Effect of maintenance

In most industrial applications, the focus is on the reliability observed on the ground, which must take into account the maintenance actions carried out. Maintenance can take several forms, depending on the level at which it is being performed (components, products, etc.). At the component level, maintenance is generally referred to as "perfect", also known as "corrective maintenance", since defective components are replaced with new ones.

At the product level, maintenance may be referred to as "preventive". This is the case with cars, for example, where engine oil, various filters, etc., are changed on a regular basis without any failures having been observed. More generally, there is "minimal" maintenance at the product level, as replacing the defective component effectively restores the reliability of the product to the level it had before the failure.

Therefore, maintenance has an important effect on product reliability, as illustrated by the following figure.

Figure 1.4. *Example of a car that has not been maintained. For a color version of this figure, see www.iste.co.uk/bayle/maturity1.zip*

For further details on the effect of maintenance on reliability and its (rather difficult) modeling, the reader is invited to refer to Rigdon and Basu (2000), Gaudoin and Ledoux (2007) and Bayle (2019).

1.4. MTBF

For most industrial applications, the objective of reliability is MTBF. There is much confusion surrounding this acronym; indeed, MTBF may signify:

– Mean time before failures:

In this case, failure instants were observed on "n" components (or products) assumed to be identical. This is equivalent to MTTF (Mean Time To Failure), as there are no maintenance actions. This can be illustrated by Figure 1.5.

– Mean time between failures:

This refers to the mean time between two consecutive failures. If there are two failures, this means there was a maintenance action, as illustrated in the following figure.

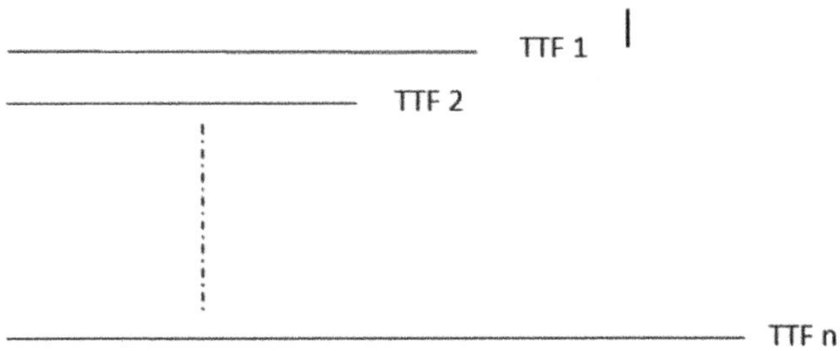

Figure 1.5. *MTBF (mean time between failures)*

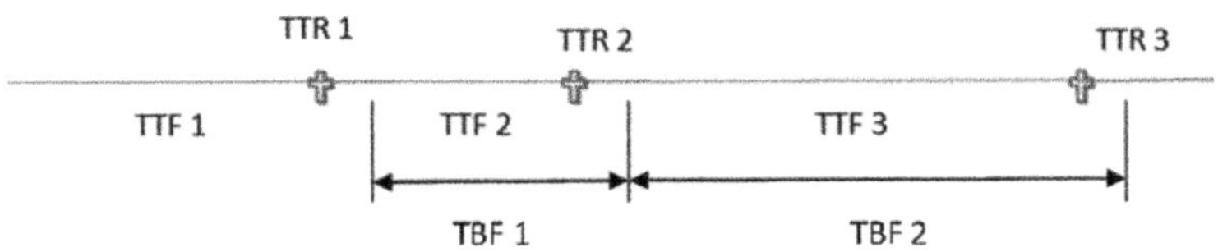

Figure 1.6. *MTBF (mean time between failures). For a color version of this figure, see www.iste.co.uk/bayle/maturity1.zip*

When there are maintenance actions, the concept of failure rate has no meaning after the first failure. Hence, time between failures (TBF) and time to repair (TTR) are used. MTBF is therefore defined here by:

$$MTBF = \mathrm{MTTF} + \mathrm{MTTR} \tag{1.8}$$

NOTE.– In practice, the TTR is often very short compared to the TBF; thus, the numerical expression of equation [1.8] can be written as:

$$MTBF \approx \mathrm{MTTF} \tag{1.9}$$

Moreover, if the product is mature (no youth or aging failure), then $\text{MTTF} = \frac{1}{\lambda}$.

According to these hypotheses, equation [1.8] can be written as:

$$MTBF \approx \frac{1}{\lambda}. \qquad [1.10]$$

This equation is often found in the literature but is only numerically true under certain hypotheses (exponential distribution), which must be verified.

1.5. Nature of the reliability objective

Product specifications always include a reliability objective. There are two main industrial applications:

– The first is less common, requiring a probability of success. This probability, which is a function of the product use time, is therefore generally provided after the product becomes operational. The unilateral lower bound of this probability is generally used as the reliability objective. This is due to the fact that it applies to one or several products for which operational failure is to be excluded (e.g. Ariane rockets or certain military weapons).

– The second covers all other applications (avionics, motor vehicles, rail, etc.) where the mean number of failures is examined. This is the well-known MTBF.

2

Maturity

2.1. Context

Any product goes through a number of industrial phases throughout its lifetime; this is known as the product lifecycle. Chronologically, these phases can be listed as follows:

– product specification, conducted by the system manufacturer or the end user;

– product design, conducted by the equipment manufacturer;

– product manufacturing, conducted by the equipment manufacturer;

– product integration, conducted by the system manufacturer;

– product operation, conducted by the end user.

These various stages are illustrated in Figure 2.1 below.

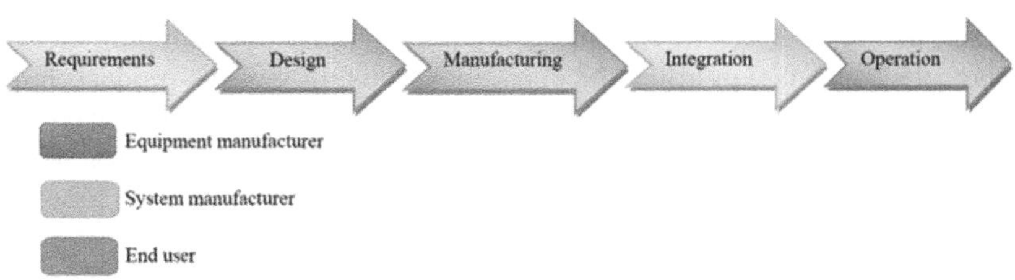

Figure 2.1. *Phases of the product lifecycle. For a color version of this figure, see www.iste.co.uk/bayle/maturity1.zip*

Chapter written by Serge ZANINOTTI.

All of these stages have the potential to produce defects that are often specific to each of them. These defects will potentially generate failures during the operational life of the product with the end user.

The specification stage runs the risk of generating:

– youth failures, as a result of failures that occur early on in the product lifecycle because a specification has been poorly met, understood or interpreted;

– premature aging failures, for example, as a result of a performance specification that is much lower than necessary for the product;

– catastrophic failures due to the lack of specifications on the operational constraints that the product will be subjected to.

The design stage runs the risk of generating:

– youth failures due to poor design of a given product function;

– premature aging failures because, for example, a component is under too high "stress" with respect to the performances guaranteed by the manufacturer;

– catastrophic failures due to the lack of margins on certain operational constraints.

The manufacturing stage runs the risk of generating:

– youth failures due to poor manufacturing of the product (poor welding, component errors, etc.);

– premature aging failures due to a component undergoing too much "stress" during product manufacturing or testing.

Finally, the integration stage runs the risk of generating catastrophic failures due to product handling errors when undergoing stress (ESD, for example), etc. All of these potential defects will affect the operational reliability of the product, having a direct impact on its maturity. But they also impact the brand image of the equipment manufacturer as perceived by the system manufacturer or the end user. They can also generate significant costs for the equipment manufacturer.

Therefore, in order to counteract these harmful effects on maturity, a certain number of analyses and tests can be implemented in order to minimize their number and therefore their impact (unfortunately, zero defect does not exist). This book relates exclusively to the level of the equipment manufacturer. Maturity can then be broken down into two large activities as depicted in Figure 2.2.

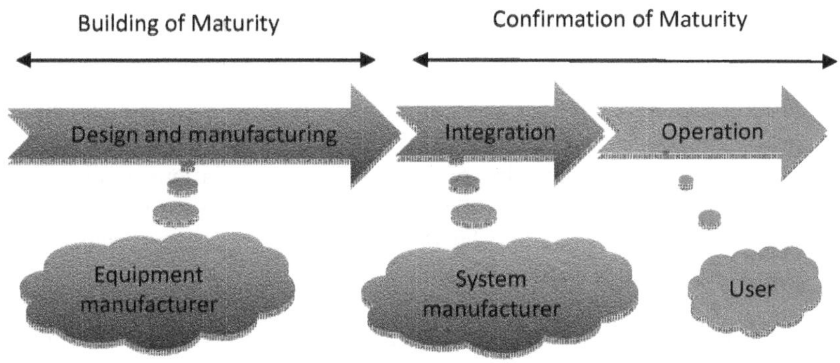

Figure 2.2. *Stages of maturity. For a color version of this figure, see www.iste.co.uk/bayle/maturity1.zip*

2.2. Normative context and its implications

2.2.1. *Quality standards*

Standards reflect the willingness of nations and individuals to exchange products or services. This trade has always required the parties' mutual trust. The legal system containing the hierarchy of standards and the quality bearing the documentary pyramid contributed to consolidating this trust over time.

This led to the creation of the world's largest standardization system in 1947, known as the International Organization for Standardization (ISO). Nowadays, to ensure their continued existence, organizations must have a certified quality management system.

ISO 9001 (ISO 2015), an international standard with worldwide coverage for several decades, sets out the criteria related to the processes that are part of the quality management systems. Having products or services offering, since the very start of operation, the ability expected by the clients is part of the objectives set through the project management process. Making sure that the project management process reaches its objectives throughout the lifecycle of the products or service is at the core of quality assurance management.

2.2.2. *Quality management system and product quality*

The introduction of a risk-based approach in ISO 9001:2015 [2] changed the way in which quality and quality assurance are managed within projects. The organizations must better understand their context and base their planning on risk determination, which requires a performance orientation of the criteria of this new version.

ISO 31000 (ISO 2018) defines the guidelines of risk management and states that it improves performance in reaching objectives. The PR-NF-EN 31010 (ISO 2018) standard supports this standard by providing recommendations concerning risk management and various techniques for taking uncertainty into account. Moreover, the ISO 9000 (ISO 2015) standard related to quality management systems, essential principles and terminology, defines the concepts used.

It provides clear definitions of objective, project, result, resource, risk, performance, efficiency, effectiveness and indicator. Consequently, the themes involved in project-based product quality management can be schematically represented by the Gilbert model.

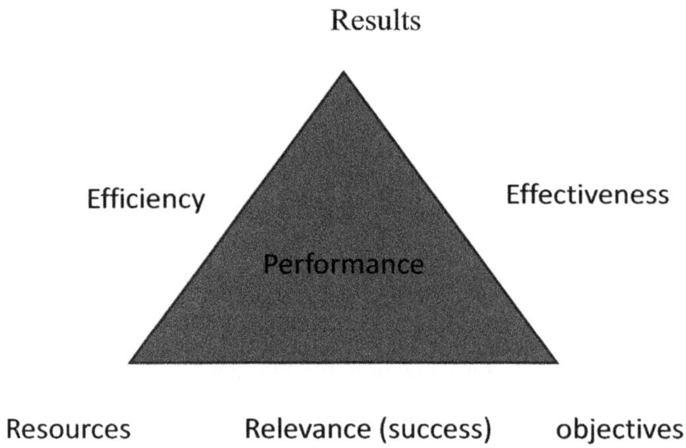

Figure 2.3. *Gilbert model*

The project objective concerns the expected product quality and meeting the cost and deadlines.

The project result concerns product quality performance, costs and deadlines.

Performance corresponds to the measurable result. Effectiveness is defined as achieving the planned activities and obtaining the planned results. Efficiency is the ratio between the achieved result and the resources used.

Success relates to a balance between economic or financial interests (resources) and the needs of the stakeholders. Risks are the effects of uncertainties. Here, the uncertainties are the causes that are susceptible to gaps with respect to the expected results.

Risk management therefore involves avoiding:

– gaps between objectives and results, affecting the project effectiveness;

– exceeding resources, affecting project efficiency.

The Ishikawa method with the "5 whys" technique is used to identify causes based on the adverse effects of a risk. The FMECA method (failure modes, effects and criticality analysis) can be used to deduce, based on the identification of risk causes, the consequences for the products in order to achieve the required improvement.

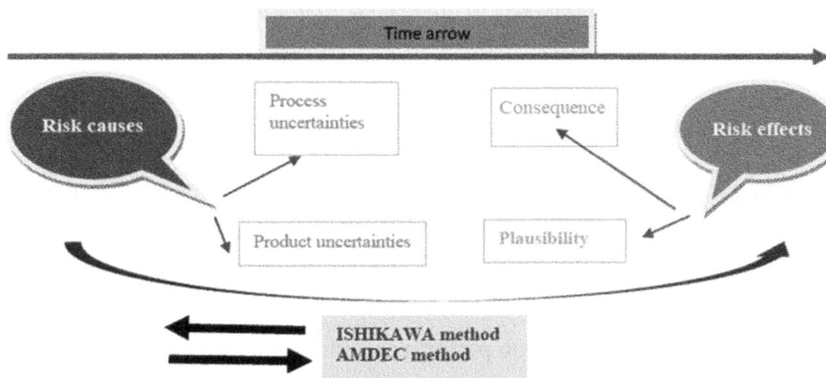

Figure 2.4. *Ishikawa method. For a color version of this figure, see www.iste.co.uk/bayle/maturity1.zip*

Successful risk management involves reaching the objectives within the allocated resources upon project completion. In this case, project performance is successful in terms of effectiveness and efficiency. It contributes to innovation by creating and preserving value. Project progress is evaluated by taking into account the identified risk factors.

The main quality objective assigned to projects is the supply of compliant products. Compliance addresses the concept of effectiveness: the product (result) must meet the client requirements (objective). Therefore, the product quality risk is essentially related to effectiveness. Indicators are used to determine, measure and control project performances. This explains the importance of their choice for product quality management.

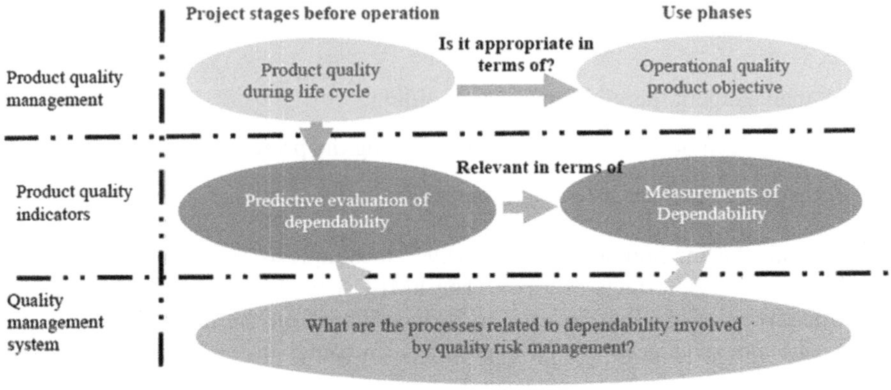

Figure 2.5. *Dependability in quality risk managment. For a color version of this figure, see www.iste.co.uk/bayle/maturity1.zip*

2.2.3. *Product quality and dependability*

According to the ISO 9000 [1] standard, the relationship between the concepts of product quality and dependability is defined as follows:

– quality designates "the ability of a set of inherent characteristics of an object to fulfill requirements";

– dependability is "the ability to perform as and when required";

– the object is "anything perceivable or conceivable", which is the product here.

ISO 9000 (ISO 2015) represents the associative relationships between quality, object and dependability as depicted in Figure 2.6.

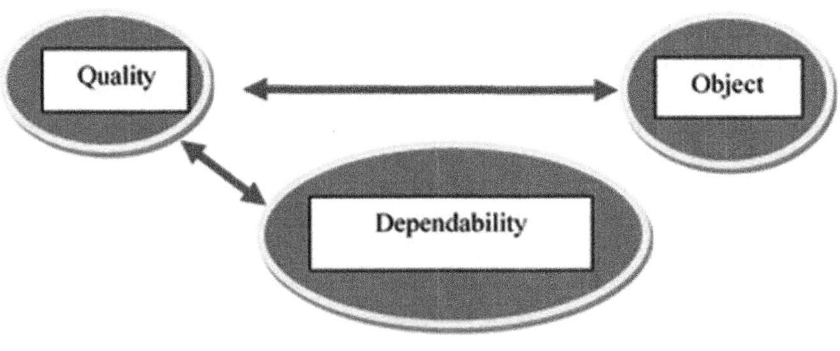

Figure 2.6. *The relationship between product quality and dependability*

Aiming to provide products that will operate perfectly throughout their operational time is completely unrealistic. The possibility that anomalies, breakdowns or failures may occur must be accepted. However, such occurrences must be limited and managed.

Managing the "operational" and "non-operational" amounts is a further way to define dependability.

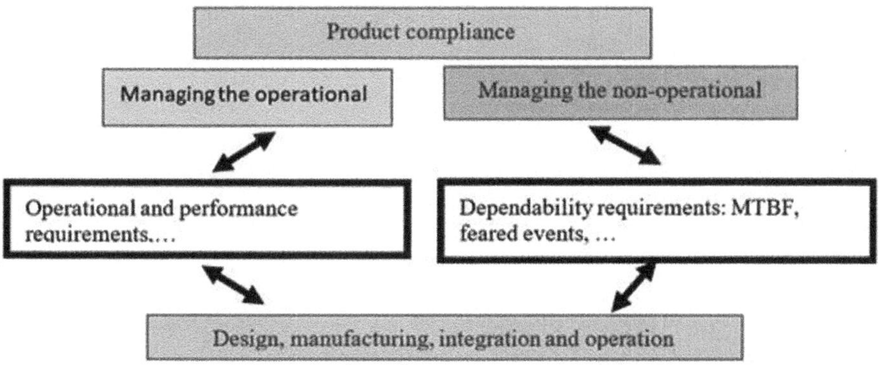

Figure 2.7. *Managing the operational and non-operational. For a color version of this figure, see www.iste.co.uk/bayle/maturity1.zip*

Dependability activities comprise two stages:

– during the operational stage, the measurement of dependability performances;

– before the operational stage, the predictive evaluation of dependability performances, which takes into account:

- the modeling of failure mechanisms,

- the existing environmental and use conditions,

- the quality of processes and activities applied throughout the lifecycle of the products,

- the inherent quality of the elementary components of the products.

The first component involves the quality control of products during their operational stage. This control is conducted by means of dependability indicators, such as the removal rate and MTBF. The second component involves a prediction of the product dependability performances that will be measured later, during the operational stage.

Before the operational stage, product compliance with the dependability requirements should be managed as a risk for the following reasons:

– there is an uncertainty related to the fact that products will comply during operation;

– this uncertainty may lead to not fulfilling the objectives related to the dependability performances.

Dependability deals with the main product quality risk, hence the interest to use it as a product quality indicator.

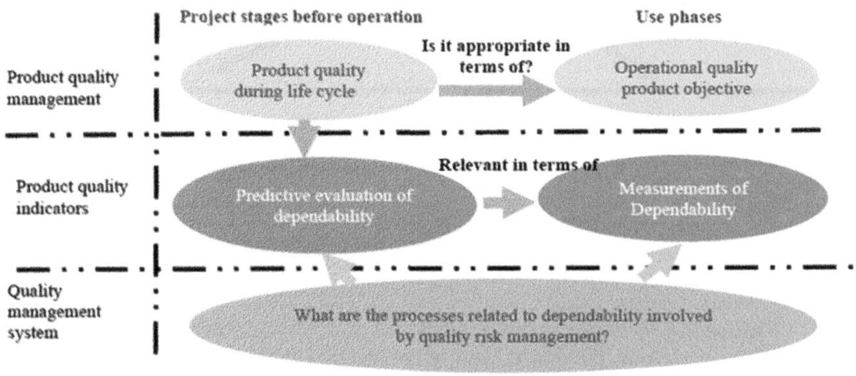

Figure 2.8. *Dependability in quality risk management. For a color version of this figure, see www.iste.co.uk/bayle/maturity1.zip*

2.2.4. *Product dependability and maturity*

2.2.4.1. *Components and indicators*

There are various preexisting standards dedicated to dependability. The European standard EN 60300-1 (IEC 2014) reproduces IEC 60300-1, referred to as ISO 9001 (ISO 2015). It states that "dependability creates trust and affects the ability of an organization to meet its objectives".

The above definition is very close to that of ISO 9000 (ISO 2015): "Dependability is the ability of an item to perform as and when required". The term "item" corresponds here to "product".

Furthermore: "Dependability includes availability, reliability, testability, maintainability and the performances of maintenance logistics and, in certain cases, other characteristics such as durability, security and safety".

Dependability commonly has four components: reliability, availability, maintainability and safety (harmlessness). A complementary component is often added, namely testability.

Reliability is the ability of an item to perform as required, without failure, for a given time interval, under given conditions. Reliability also designates the value of reliability, in which case it is the probability of a given item to fulfill a required function, under given conditions, during a given time interval.

Maintainability is the ability of an item to be retained in or restored to a state to perform as required, under given conditions of use and maintenance. When maintenance is performed under given conditions and with specified processes and means, it is part of logistical support. Maintainability is characterized by the probability of successfully achieving maintenance or by the duration and cost required to achieve maintenance.

Availability is the ability of an item to be in a state to perform as required. Availability also designates the value of availability. The availability of a product depends on its *reliability* and *maintainability*. Indeed, a product is able to perform at a given instant provided that it has either not stopped performing (*reliability*) or could be restored to a state to perform in case of failure (*maintainability*).

Safety (harmlessness) is the ability of a product to meet, throughout its life stages, an acceptable level of risk of accidents susceptible to harm personnel or seriously damage the product or its environment. Safety is characterized by the probability of an unacceptable accident to occur within a given time interval.

Testability is the ability of a system to be declared defective within specified time limits, according to specified processes and under given conditions. The *testability* of a component contributes to the *maintainability* of the equipment it contains. It is characterized by the coverage rate that is defined by the number of testable breakdowns to the total potential number of breakdowns.

Dependability performances depend on the quality of processes and activities applied during the lifecycle of products. They are defined according to the indicators in Figure 2.8. Note, for example, the proper functioning rate concerning availability, the rate of occurrence of dreaded events related to safety, the mean time between failures (MTBF) and the removal rate for reliability.

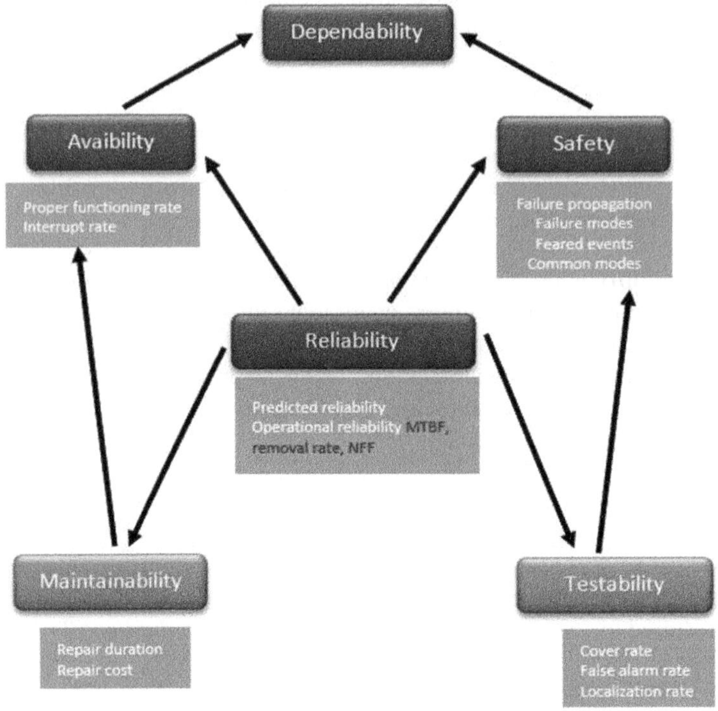

Figure 2.9. *Dependability. For a color version of this figure, see www.iste.co.uk/bayle/maturity1.zip*

2.2.4.2. *Product maturity*

The notion of product maturity is not defined by ISO 9000. Nevertheless, a measurement system is employed for the evaluation of the maturity level of a technology within the innovation framework covered by the NF ISO 16290 standard.

It was first developed and used by NASA, but its application has been extended to all domains. The scale has nine levels of maturity relying on qualification criteria. Several standards refer to the maturity of organizations and processes. NF EN ISO 9004 (ISO 2018) gives guidelines for organizations to achieve sustained success. It states: "an organization is considered mature if it performs in an effective, efficient and sustainable manner". The maturity of the organization is established by qualitative evaluation of the maturity of each of the individual processes following five levels".

The ISO/IEC 33001 standard provides complementary elements concerning software, systems and services. The internationally known Capability Maturity Model Integration CMMi (Capability Maturity Model Integration (CMMI) 2018) also involves five levels for the evaluation of the maturity of processes and organizations.

Nowadays, product maturity is a notion used by various organizations. It is sometimes defined according to the reliability component:

> State of a product that meets the expected reliability level as soon as it becomes operational.

Sometimes it is defined according to the availability component:

> State of a product that meets the availability requirement as soon as it becomes operational.

The definitions are very similar and they relate to dependability. In recent decades, manufacturers were interested in accelerated tests, at the core of the building and confirmation of product maturity, and CNES RNC-CNES-Q-20-523 can be quoted:

> Reliability clauses followed the evolutions of an industrial context in which specialization became a generalized trend: certain companies specialized in design or manufacturing, or in product assembly for a contract giver. At present, these clauses do not uniquely relate to objectives in terms of MTBF, but rather to the probabilities of failure upon the first energizing, or to the number of defects tolerable throughout the youth period, or still to the durability criteria. Moreover, these objectives must be met from the very beginning of the commercial life of the product. To meet these objectives, the manufacturers are increasingly required to respond in terms of robustness, maturity of products and their manufacturing processes (RNC 2006).

Therefore the product maturity strategy refers to all of the activities contributing to meeting the dependability objectives. Moreover, its objective is to make sure that product reliability is under control as soon as it becomes operational. The state of maturity of products therefore depends on youth and end of life according to the bathtub curve used in reliability.

The product maturity strategy is defined according to a process that is referred to here as “building and confirmation of maturity”.

The period referred to as “maturity” corresponds to the area where product reliability and therefore quality are controlled under operation. It corresponds to the period when failure intensity is constant in time, modeled by an exponential law where λ is the failure rate. When certain elements are subjected to aging and preventive maintenance, reliability is controlled by failure intensity.

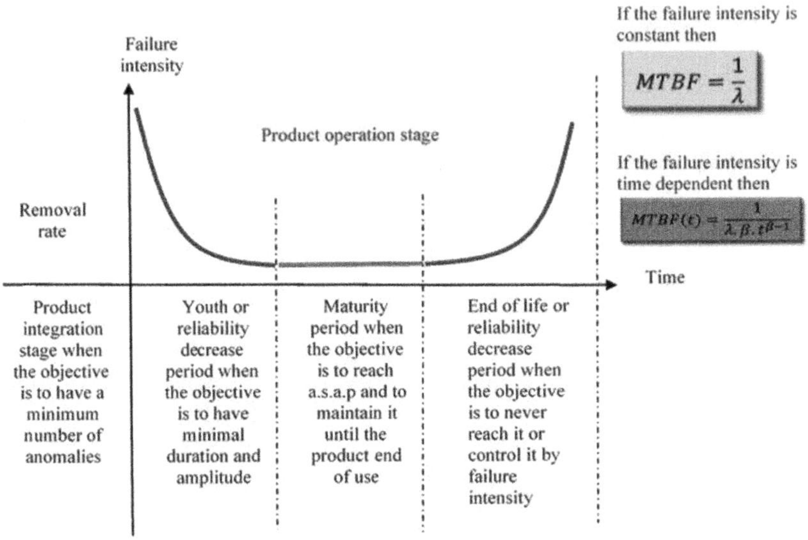

Figure 2.10. *Product operation stage. For a color version of this figure, see www.iste.co.uk/bayle/maturity1.zip*

This leads to the following representation of the relationship between product quality and product maturity.

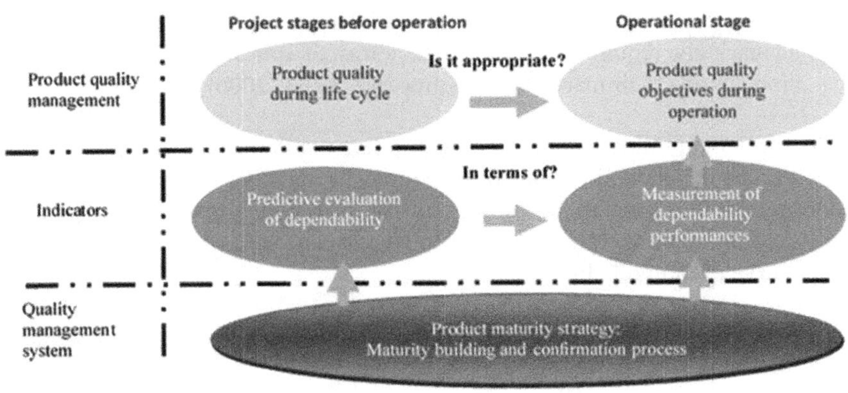

Figure 2.11. *Maturity building and confirmation process. For a color version of this figure, see www.iste.co.uk/bayle/maturity1.zip*

2.2.4.3. *PDCA and product maturity*

The processes belonging to a quality management system follow the PDCA cycle according to ISO 9001 (ISO 2015).

[P]: Plan. [D]: Do. [C]: Check. [A]: Act

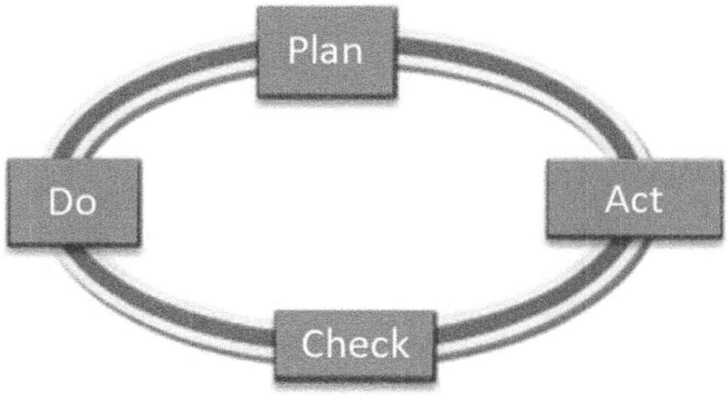

Figure 2.12. *PDCA. For a color version of this figure, see www.iste.co.uk/bayle/maturity1.zip*

The maturity building and confirmation process involves the following stages:

– planning (P) the product maturity strategy relying on the maturity building and confirmation process;

– doing (D) activities throughout the lifecycles of the products involved in the plans, without ignoring those concerning the predictive evaluations of dependability;

– checking (C) dependability performances during integration and operation, without ignoring their comparisons to predictive evaluations;

– acting for the improvement (A) of maturity building and confirmation processes according to the feedback.

2.2.5. *Standards in various domains*

The international system ISO applies to all organizations. Various domains such as the medical, railway, nuclear, space, aeronautics, defense and automotive sectors completed the ISO 9001 standard according to their specificities.

Dependability challenges are part of the main criteria involved in the complementary contributions. In the automotive field, there is a methodology dedicated to the risk-based approach to product quality management, APQP-PPAP.

It aims to optimize product quality and the associated costs by anticipating activities depending on the identified risks. In the aeronautics, space and defense fields, this approach was resumed in the form of a standard, namely PR NF EN 9145. Its objective is to provide safe and reliable products as soon as possible.

Nowadays, it can be used by any manufacturer with the suitable adjustments for best taking into account risk management.

2.2.6. *Perspectives*

2.2.6.1. *Opportunity*

Reliability is the dependability component that is connected with all of the others; see Figure 2.7. Therefore, the predictive evaluations of dependability mainly rely on reliability and failure rate control. In recent decades, the quality and reliability fields became highly specialized. Nowadays, quality engineers do not always consider using the dependability evaluations and measurements as a tool for product quality management.

Reliability engineers do not consider cooperating with quality engineers for the validation of process quality hypotheses formulated in the dependability evaluations. However, during the same period, new methods for predictive evaluations of reliability were developed, such as FIDES by the Risk Management Institute and 217Plus by RIAC (2015).

They use process quality factors, referred to as pi-process, evaluated through audit, as recommended by the FIDES guide. These factors aim to evaluate the quality level of the activities performed throughout the product lifecycle, and to quantify the impact on the values of failure rates and MTBF.

The calculation of failure rates, also referred to as risk rates, takes into account all of the activities of maturity building and confirmation associated with the set of dependability indicators.

On this subject, the FIDES guide notes that the audit is a tool for risk management concerning particularly:

– the robustness of product definition;

– the environment of operational products;

– the effective consideration of reliability throughout the lifecycle.

The FIDES audit can be used in addition to the ISO 9001 V2000 audit, which is more specific and dependability oriented. The specific objectives of conducting a FIDES audit are:

– evaluation of a quantified quality indicator that is representative of the capacity of the company to manage the reliability of its products (pi-process);

– evaluation of a reliability sizing factor: pi-process calculation;

– identification of the company strengths and weaknesses and formulation of targeted recommendations for the improvement of the internal process.

To the extent that product quality management, predictive evaluation and measurement of dependability performances are jointly conducted, it is possible to build quantified indicators related to product quality risks.

2.2.6.2. *Building a quantified product maturity indicator*

Projects must have relevant indicators to manage quality risks. Indicators established according to quantitative criteria are preferable; they are more robust and reliable than the qualitative ones, whose evaluations depend on those conducting them.

Failure rate and MTBF indicators can be used to quantify the quality risks during operation and compare them with requirements, as shown in Figure 2.8. But they are not effective in managing quality risks in the upstream stages, before the delivery of products.

Thanks to the pi-process of the FIDES and 217Plus methods, the evaluation of the quality level of the activities applied throughout the lifecycle of products can be used to quantify the impact on predictive reliability, the image of quality risks perceived by clients during operation stage.

At the start of the project, the planning stage [P] defines the activities involved by the maturity building and confirmation process. In the subsequent stages, activities are conducted and evaluated [D]. Pi-process-based evaluations after each activity offer the possibility of real-time updates on the value of predictive MTBF.

During these stages, the risk-based approach involves the adjustment of predictive MTBF so that it is higher than the objective MTBF by means of the choice and/or level of quality of activities. The measurement of operational MTBF can be used to calibrate the predictive reliability model [C] and improve the maturity of the process of maturity building and confirmation [A]. During this operational stage, the risk-based approach involves maintaining the functionality of the products and checking that the operational MTBF is always higher than the objective MTBF.

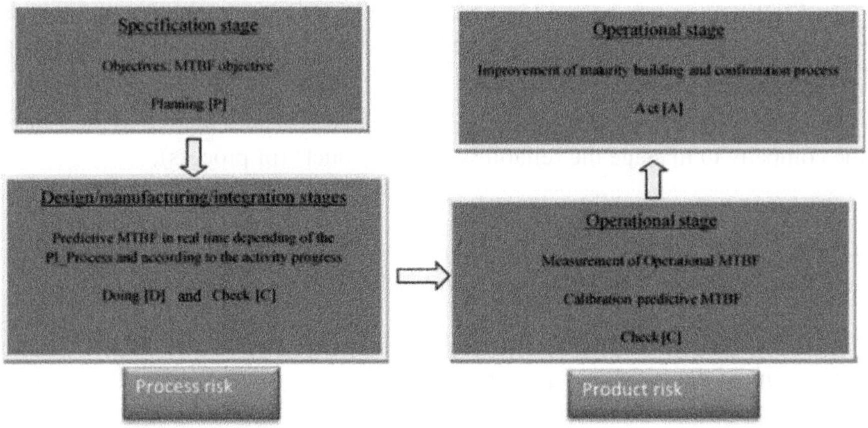

Figure 2.13. *Process and product risk. For a color version of this figure, see www.iste.co.uk/bayle/maturity1.zip*

As previously noted, process maturity is related to effective and efficient functioning objectives. Product maturity is related to the objective of reaching the reliability level as soon as the product is operational. Quality indicators are essentially related to effectiveness. It is however possible to create product maturity indicators that are also related to efficiency.

As previously explained, the risks to be managed before and after the operational stage are not the same. The operational MTBF considered here is the one measured when the product becomes operational. This is how the youth period of the bathtub curve is taken into account.

2.2.6.2.1. Process risk quantified indicator

Before the operational stage, the value of operational MTBF cannot be measured. Quality risk concerns the activities of the maturity building and confirmation process. The risk is to apply a maturity building and confirmation process that does not allow the objective MTBF to be reached.

The value of predictive MTBF can be compared to that of the objective MTBF. This amounts to managing process risks using the $R_{process}$ indicator.

Given:

$$R_{process} = 1 + \frac{MTBF_{objective} - MTBF_{predictive}}{MTBF_{objective}}$$

– If $R_{process}$ > 1, the process is not sufficiently effective, the quality level must be improved or the choice of activities must be reviewed. See Figure 2.1.

– If $R_{process}$ < 1, the process is sufficiently effective.

– If $R_{process}$<< 1, the process is effective, but it is all the less efficient as the [R process] gets farther from 1. There is an opportunity for limiting the resources. See Figure 2.1.

2.2.6.2.2. Product maturity quantified indicator

In the operational stage, the value of operational MTBF is measurable. The risks are related to the product. Product risks are managed by the product maturity indicator: M product. The risk is to have failures more frequently than the objective expected by the client.

$$M_{product} = 1 + \frac{(\mathrm{MTBF}_{\mathrm{objective}} - \mathrm{MTBF}_{\mathrm{operational}})}{\mathrm{MTBF}_{\mathrm{objective}}}$$

– If $M_{product}$ > 1, the product is not sufficiently mature.

– If $M_{product}$< 1, the product is sufficiently mature.

– If $M_{product}$<< 1, the product is significantly more mature than required. A value analysis should be conducted.

2.2.6.2.3. Process maturity quantified indicator

The maturity building and confirmation process is effective if the predictive MTBF calculated by taking into account the choice and quality level of the applied activities is higher than the operational MTBF. The maturity building and confirmation process is all the more efficient as the predictive MTBF is closer to operational MTBF. In this case, the deployed resources are up to the objective. The maturity level of the maturity building and confirmation process can be evaluated based on the following indicator:

$$M_{process} = 1 + \frac{(\mathrm{MTBF}_{\mathrm{predictive}} - \mathrm{MTBF}_{\mathrm{operational}})}{\mathrm{MTBF}_{\mathrm{operational}}}$$

– If $M_{process}$ > 1, the process is sufficiently effective.

– If $M_{process}$ < 1, the process is not sufficiently effective, the choice must be reviewed and/or the quality of the applied activities must be improved. The process is not sufficiently mature.

– If $M_{process}$ >> 1, the process is effective, but all the less efficient as the [M process] gets farther from 1. There is an opportunity for limiting the resources.

– If $M_{process}$ > 1 but close to 1, the process is effective and efficient, it is mature.

– If the calibration of predictive reliability models can be used to calibrate the process quality contributor, here referred to as the pi-process, it is possible to transpose these indicators defined according to MTBF values using the pi-process corresponding values.

2.3. Building of maturity

The aim is to implement actions that ensure product maturity before its operation by the end user. Therefore, for the design stage, the following analyses are proposed:

– derating: this analysis is the subject of Chapter 4;

– components with limited life duration: this analysis will be the subject of Chapter 5;

– worst-case analysis: this analysis will be the subject of Chapter 6.

For the manufacturing stage, the following tests are proposed:

– robustness test or accelerated test: this test will be the subject of Chapter 7;

– burn-in test: this test will be the subject of Chapter 8;

– run-in test: this test will be the subject of Chapter 2 of a second volume by the author, *Product Maturity 2: Principles and Illustrations*, hereinafter referred to as Volume 2;

– manufacturing sampling: this method will be the subject of Chapter 3 of Volume 2;

– compliance test: this test will be the subject of Chapter 4 of Volume 2;

– non-regression test: this test will be the subject of Chapter 5 of Volume 2;

– reliability proof: this test will be the subject of Chapter 6 of Volume 2;

– reliability management: this analysis will be the subject of Chapter 7 of Volume 2;

– confirmation of maturity: this analysis will be the subject of Chapter 8 of Volume 2.

According to Figure 2.8, the stage of maturity building is conducted by the equipment manufacturer, prior to any delivery to the system manufacturer, during the design and manufacturing stages. The aim of this activity is to limit and even eliminate the defects generated during the two stages. A certain number of tests can be conducted for this purpose. For the defects in the manufacturing stage, burn-in tests can be conducted on all of the products before their delivery. This will be the subject of Chapter 8.

For the design stage, besides the qualification tests specified by the system manufacturer, robustness tests can be conducted. This robustness can be assessed by tests referred to as accelerated. Figure 2.14 illustrates the correspondence between various types of tests.

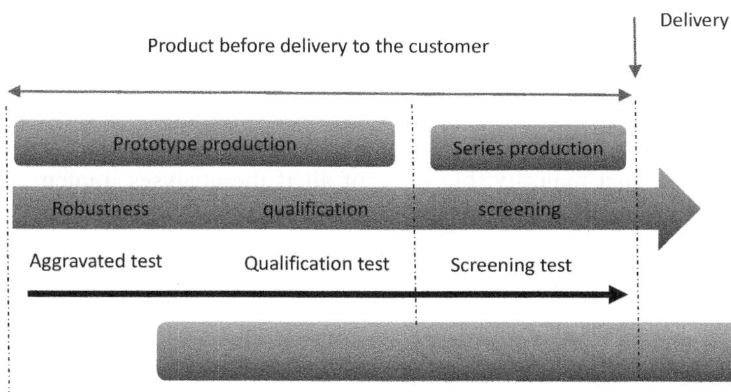

Figure 2.14. *Tests during the manufacturing stage. For a color version of this figure, see www.iste.co.uk/bayle/maturity1.zip*

NOTE.– The qualification stage is often contractual and relies on a standard. The qualification tests aim to check that the product is able to function within a specified environment before becoming operational. Time is not involved, so they cannot be used to evaluate product reliability.

Product maturity is built based on a certain number of analyses, rules, etc. touching all stages of the lifecycle of the product. Their objective is, among others, to reach the required product reliability and consequently, the various failure categories.

Youth failures, though rare and with no major impact on reliability (MTBF), have a direct impact on the brand image for the end user. The inconvenience resulting from a failure occurring within several days of buying a car, domestic appliance or any other product can easily be understood. The other important point is that youth failures can have financial impacts when they are observed too rapidly. This can take the form of a warranty (automobile, domestic appliance, etc.) or financial penalties based on the removal rate at the equipment manufacturer.

It is a known fact that, according to the second law of thermodynamics, everything degrades over time. Therefore, any component is subject to aging phenomena. These should be taken very seriously, as they concern the entire population of operational products. The aim should therefore be, particularly during the design stage, to avoid the occurrence of this type of failure throughout the operational life of the product. As for the catastrophic failures, they may be so numerous that the required level of reliability cannot be met. Their occurrence frequency can be reduced by obtaining robustness margins, worst-case analyses, etc.

2.4. Confirmation of maturity

Product maturity is obviously measured during the operational stage. This aims to check the product maturity, but most of all if the analyses implemented in the stage of maturity building are effective. For example:

– the detection of too high composed reliability: this analysis will be the subject of Chapter 7 of Volume 2;

– the estimation of the operational reliability of the product: this analysis will be the subject of Chapter 8 of Volume 2.

In order to make sure that the product is actually mature (exempt from youth defects and properly designed to avoid rapid wear), it is important to confirm its maturity. Maturity confirmation is conducted in two stages:

– The first stage is conducted by the system manufacturer by measuring the removal rate, the original fit rate or 0 defects. Quite often, the system manufacturer provides an objective removal rate with pending penalties, if the objective is not met.

– The second stage, involving the evaluation of the effectiveness of burn-in and accelerated tests, is conducted by the equipment manufacturer based on the operation data provided by the system manufacturer and the end user.

This very important stage of maturity confirmation is illustrated in Figure 2.15.

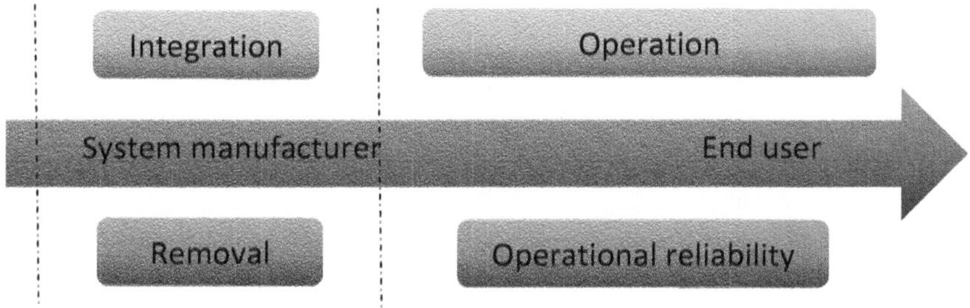

Figure 2.15. *Confirmation of maturity. For a color version of this figure, see www.iste.co.uk/bayle/maturity1.zip*

3

Derating Analysis

3.1. Derating

Derating is conducted according to rules that preserve a "safety margin" (robustness) in terms of the maximum values of components guaranteed by manufacturers, in the case of operational constraints (electric overloads, disruptions, overheating, etc.) or in the case of errors or accidents while in use.

From a theoretical viewpoint, derating is defined by the following equation:

$$Derating = \frac{\text{Applied value}}{\text{Guaranteed maximum value}} \quad [3.1]$$

Let us illustrate this by considering the example of a capacitor whose manufacturer guarantees a maximum voltage of 50 V. If, in the application concerned, a voltage of 20 V is applied, the derating will be 20/50 = 40%. In order to check if this derating is acceptable, the derating rules must be clarified.

Derating has a positive impact on the reliability of components. Consider the three types of failures mentioned in Chapter 1. It is clear that derating has little to no influence on youth failures because they are generally the result of manufacturing errors.

However, derating has some influence on catastrophic failures. It is important to recall that the latter are normally the result of accidental overloads, for example, overvoltage. If we return to the example of the capacitor mentioned above, we see that, in certain cases, derating may prevent a failure from occurring if the value of the overvoltage remains below the guaranteed maximum voltage. Similarly, a slightly undersized component can lead to premature aging. The derating applied may then push this aging beyond the point when the product integrating the component is in operation.

There are several data sources that indicate the derating rules to be applied. Some of them are listed below:

– Certain rules indicated by the manufacturers of components are detailed in section 3.2.

– Rules outlined in derating standards are listed in section 3.3.

To our knowledge, there is no justification for the rules proposed in the standards and fairly often these rules continue to be applied from one generation of products to the next, while component technology continues to evolve. Furthermore, these rules do not seem to be established according to physical laws of failures that model the component failure mechanisms. Finally, these rules are sometimes different from one standard to another, most likely due to the domain of the product application (aeronautics, rail, space, automotive, etc.).

On the basis of these findings, section 3.3 proposes a method for minimizing these defects and, above all, provides justification for the derating rules under consideration.

3.2. Rules provided by the manufacturers of components

3.2.1. *CMS resistors*

For this type of component, there are three parameters for which the maximum value guaranteed by the manufacturer must not be exceeded.

3.2.1.1. *Dissipated power*

3.2.1.1.1. Steady state

The maximum dissipated power depends on the type of enclosure and on the ambient temperature. It is generally provided in the datasheet for an ambient temperature between -55°C and +70°C. Taking the example of Vishay resistors (Vishay 2018), this power decreases linearly above +70°C until it becomes zero at a temperature around 155°C, as illustrated in the figure below (Vishay 2018).

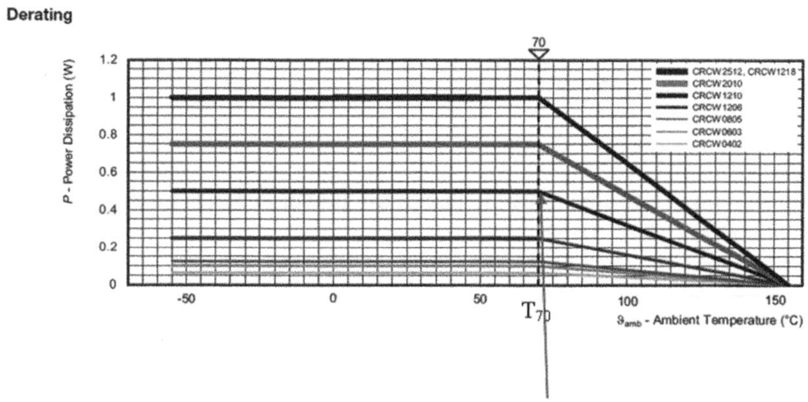

Figure 3.1. *Power derating according to the temperature of a CMS resistor*

For example, a 1210 resistor can dissipate a maximum power of 500 mW, up to an ambient temperature of 70°C. The table below indicates the maximum power by the type of CMS resistor.

Resistor type	Maximum direct power (mW)
0402	63
0603	100
0805	125
1206	250
1210	500
1218	1,000
2010	500
2512	1,000

Table 3.1. *Power dissipated by each type of resistor*

The temperature derating curve can be modeled as follows:

$$Pr(Ta) = Pmax\,.\left[\beth_{Ta\leq T_{70}} + \beth_{Ta>T_{70}}\cdot\left(\frac{Tmax-Ta}{Tmax-T_{70}}\right)\right] \quad [3.2]$$

For example, for a 0805 enclosure and an ambient temperature of 85°C, the following data are obtained:

$$\text{Ta} = 85°\text{C} \quad \text{Pmax} = 125 \text{ mW; therefore, } \text{Pr}(85°\text{C}) = 125 \,.\left[0 + 1.\left(\frac{155-85}{155-70}\right)\right]$$

or $\text{Pr}(85°\text{C}) \sim 103 \text{ mW}$

3.2.1.1.2. Transient state

For transient power dissipations, the maximum power depends on the duration of this transient state (Vishay 2018).

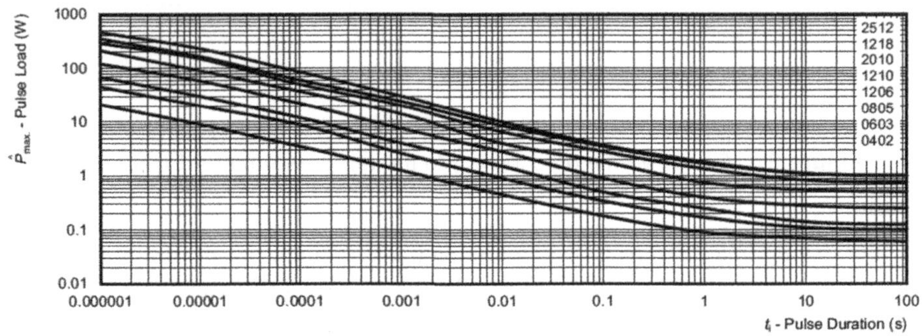

Figure 3.2. *Power derating for a Vishay CMS resistor under single pulse state*

For repetitive pulses, Figure 3.3 is obtained (Vishay 2018).

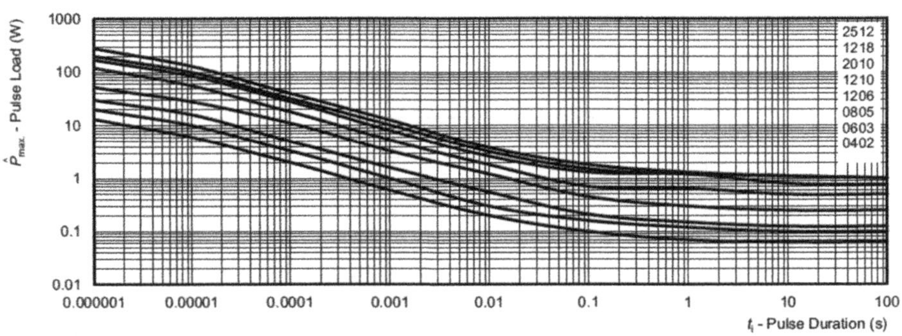

Figure 3.3. *Power derating for a Vishay CMS resistor under repetitive pulses*

For example, a 1206 resistor can dissipate a maximum power of 300 mW for 1 s.

3.2.1.2. *Applied voltage*

3.2.1.2.1. Steady state

Under a steady state, the voltage across resistors must not exceed the maximum value guaranteed by the manufacturer. These values are given in the following table.

Type of resistors	Maximum voltage (V)
0402	50
0603	75
0805	150
1206	200
1210	200
1218	200
2010	400
2512	500

Table 3.2. *Voltage depending on the type of resistor*

3.2.1.2.2. Transient state

In a transient state, the voltage applied may be higher than in a continuous state.

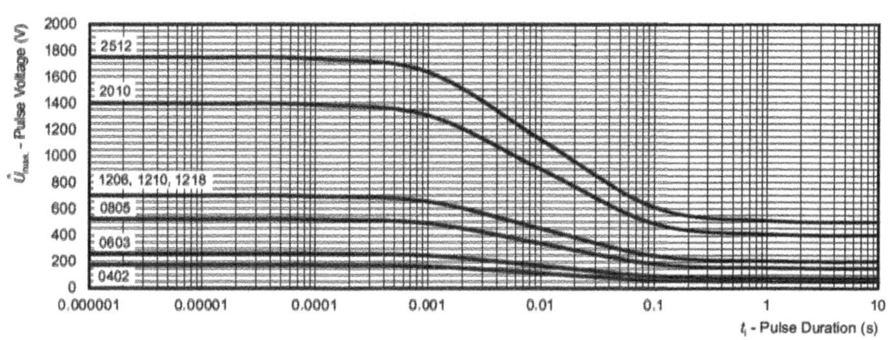

Figure 3.4. *Voltage derating for a Vishay CMS resistor*

For example, for a 1206 resistor, a maximum voltage of ~500 V can be applied for 10 ms.

3.2.2. *Capacitors*

In this section, we will discuss ceramic capacitors (types I and II), solid tantalum capacitors and electrolytic capacitors.

3.2.2.1. *Ceramic capacitors*

The parameter of which the maximum value should not be exceeded is the voltage across the capacitor. This voltage depends on the type of enclosure. Generally, there is no derating mentioned as a function of temperature.

However, certain manufacturers specify (but do not always guarantee) a lifetime under specific conditions (derating), as illustrated below (Murata 2020).

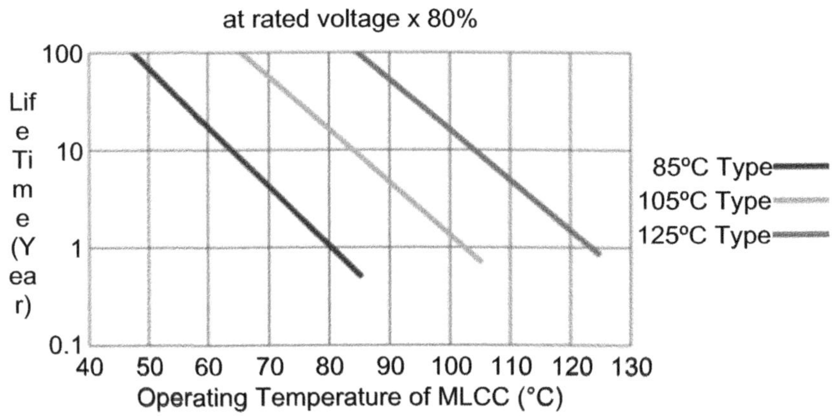

Figure 3.5. *Lifetime of an MLCC (multilayer ceramic capacitor) as a function of temperature. For a color version of this figure, see www.iste.co.uk/bayle/maturity1.zip*

NOTE.– For this manufacturer, lifetime is defined by the time elapsed until 1% of the components fail;

– since the three lines are parallel, the activation energy is the same for the three classes of temperature.

3.2.2.2. *Tantalum capacitors*

For this type of component, there are three parameters for which the maximum value guaranteed by the manufacturer should not be exceeded.

3.2.2.2.1. Direct voltage

Direct voltage is the voltage across the capacitor. It should not exceed the maximum value UR guaranteed by the manufacturer. A temperature derating is provided by the manufacturer (Firadec 2015).

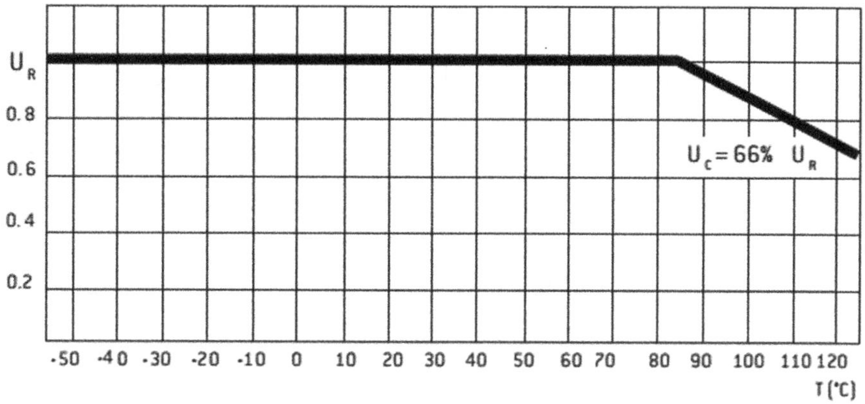

Figure 3.6. *Temperature derating of voltage across a tantalum capacitor*

This temperature derating curve can be modeled as follows:

$$V(Ta) = V_R \,.\left[\beth_{Ta \le T_{85}} + \beth_{Ta > T_{85}} \cdot \frac{1}{3} \cdot \left(\frac{Tmax - Ta}{Tmax - T_{85}}\right)\right] \qquad [3.3]$$

with Tmax = 125°C and T85 = 85°C.

3.2.2.2.2. Dissipated power

This is presented in Figure 3.7 (Firadec 2015).

This temperature derating can be modeled as follows:

$$P(Ta) = Pmax \,.\left[\beth_{Ta \le T_{25}} + \beth_{Ta > T_{25}} \cdot a.\left(1 - b.exp(c.Ta)\right)\right] \qquad [3.4]$$

with a = 1.1279; b = 0.086395; c = 0.018611.

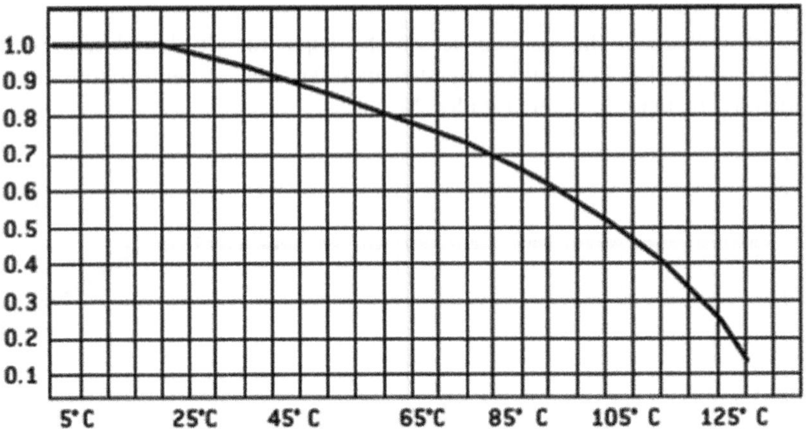

Figure 3.7. *Temperature derating of the power dissipated by a tantalum capacitor*

3.2.2.3. *Electrolytic capacitors*

Four stresses should be considered for ëlectrolytic capacitors:

– The direct voltage applied, equal to the continuous voltage plus the positive peak voltage due to the equivalent series resistance (ESR) of the capacitor.

– Overvoltage for short periods, which should not exceed 1.15 times the rated voltage guaranteed by the manufacturer.

– The reverse voltage applied that must be below or equal to 1 V.

– Storage duration. There is no variation in electrical characteristics when stored for long periods of time, but leakage current may increase. This is due to chemical reactions between the alumina dielectric and the electrolyte. These reactions are reversible when the capacitor is under voltage. The manufacturer normally guarantees storage duration without reformation under well-defined temperature conditions and durations, for example, up to temperatures of 50°C under the following conditions:

- Voltage below 100 V ➔ 5 years.
- Between 100 V and 360 V ➔ 3 years.
- Between 360 V and 500 V ➔ 1 year.
- Above 500 V ➔ 6 months.

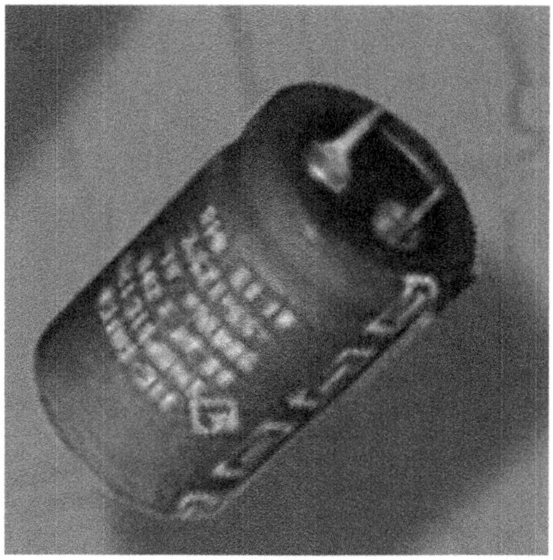

Figure 3.8. *Example of electrolytic capacitor*

Similar to with tantalum capacitors, it is recommended to have voltage derating depending on the operating temperature. A 60% derating is often recommended.

NOTE.– This type of capacitor is mainly used as an energy reserve (compensating for the power supply microcuts). Due to their large overall external dimensions, a 60% derating rate, as mentioned in the "derating summary" table (Table 3.6), is not achievable at the industrial scale. Moreover, a more significant derating is often considered as going up to 90%.

3.2.3. *Magnetic circuits*

The parameters to be monitored are:

– maximum temperature;

– saturation current.

Generally, there are no derating rules imposed by the manufacturers.

3.2.4. *Fuses*

The main stresses are:

– the current through the fuse (breaking capacity);

– operating temperature;

– voltage across the fuse when it is "open".

Generally, there are no derating rules imposed by the manufacturers.

3.2.5. *Resonators*

The main stress is the operating temperature. Generally, there are no derating rules imposed by the manufacturers.

3.2.6. *Oscillators*

The main stresses are:

– operating voltage;

– supply voltage.

Generally, there are no derating rules imposed by the manufacturers.

3.2.7. *Photocouplers*

There are many stresses to be considered:

– photodiode reverse voltage;

– photodiode reverse peak voltage;

– photodiode direct current;

– maximum power dissipated in the photodiode;

– breakdown voltage VCEo of the phototransistor;

– breakdown voltage VCBo of the phototransistor;

– breakdown voltage VBEo of the phototransistor;

– maximum power dissipated in the phototransistor;

– maximum operating temperature.

Generally, there are no derating rules imposed by the manufacturers.

3.2.8. *Diodes*

The main stresses are:

– maximum direct current If;

– maximum repetitive peak current IFRm;

– maximum continuous reverse voltage Vr;

– maximum repetitive peak reverse voltage VRRm;

– maximum junction temperature;

– maximum dissipated power.

Generally, there are no derating rules imposed by the manufacturers.

3.2.9. *Zener diodes*

The main stresses are:

– maximum dissipated power;

– maximum junction temperature.

Generally, there are no derating rules imposed by the manufacturers.

3.2.10. *Tranzorb diodes*

The main stresses are:

– maximum pulse power Pppm;

– maximum operating temperature;

– steady-state power.

3.2.10.1. *Maximum pulse power*

Maximum pulse power is derated depending on pulse duration. The curve below, derived from the manufacturer data, is defined for a non-repetitive bi-exponential wave. This is generally how lightning pulses are described in the specifications of the product concerned.

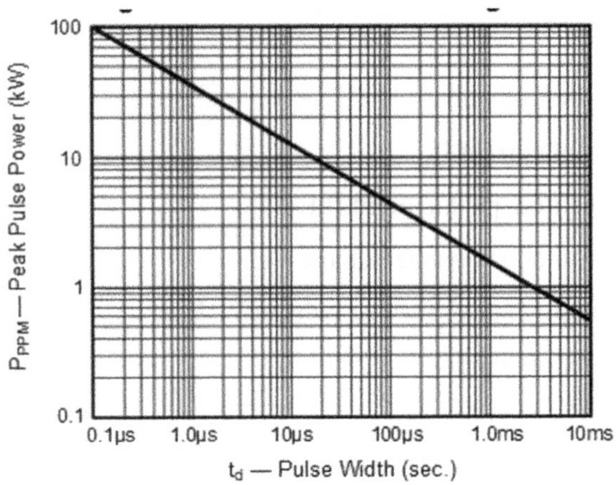

Figure 3.9. *Acceptable maximum peak power in a Tranzorb diode*

3.2.10.2. *Steady-state power*

Generally, the manufacturer provides a temperature derating of the steady-state dissipated power, as illustrated below.

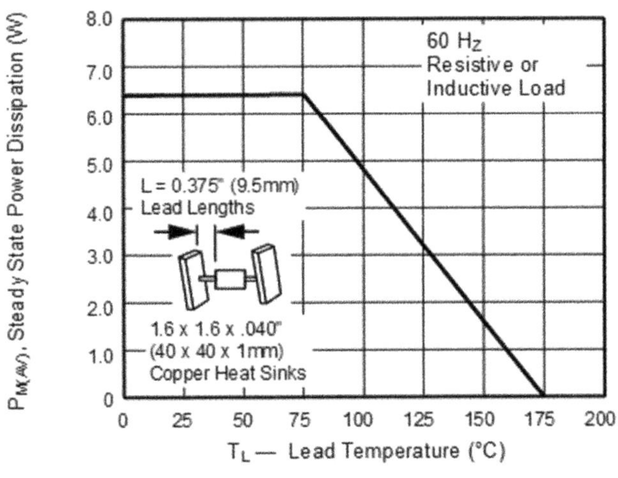

Figure 3.10. *Temperature derating of the steady-state dissipated power*

3.2.10.3. *Pulse power*

Generally, the manufacturer provides a temperature derating of the steady-state dissipated power, as illustrated below.

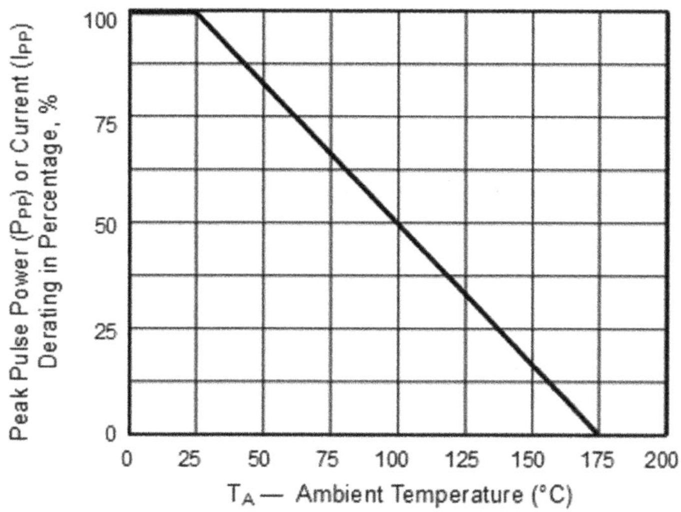

Figure 3.11. *Temperature derating of the dissipated pulse power*

There is no derating for the junction temperature.

3.2.11. *Low power bipolar transistors*

The main stresses are:

– collector emitter voltage VCEo;

– emitter base voltage VBE;

– collector direct current;

– junction temperature.

Generally, there are no derating rules imposed by the manufacturers.

3.2.12. *Power bipolar transistors*

The main stresses are:

– collector emitter voltage VCEo;

– emitter base voltage VBE;

– permanent collector direct current;

– transient collector direct current;

– base current;

– junction temperature.

Generally, there are no derating rules imposed by the manufacturers.

3.2.13. *Low power MOSFET transistors*

The main stresses are:

– drain source voltage Vds;

– grid source voltage Vgs;

– grid drain voltage Vgd;

– steady-state direct current ID;

– pulsed state direct current ID;

– maximum junction temperature;

– maximum dissipated power.

Generally, there are no derating rules imposed by the manufacturers.

3.2.14. *High power MOSFET transistors*

The main stresses are:

– drain source voltage Vds;

– grid source voltage Vgs;

– grid drain voltage Vgd;

– steady-state direct current ID;

– pulsed state direct current ID;

– junction temperature;

– steady-state dissipated power.

There is generally a derating of the dissipated power depending on the case temperature Tc, as illustrated below.

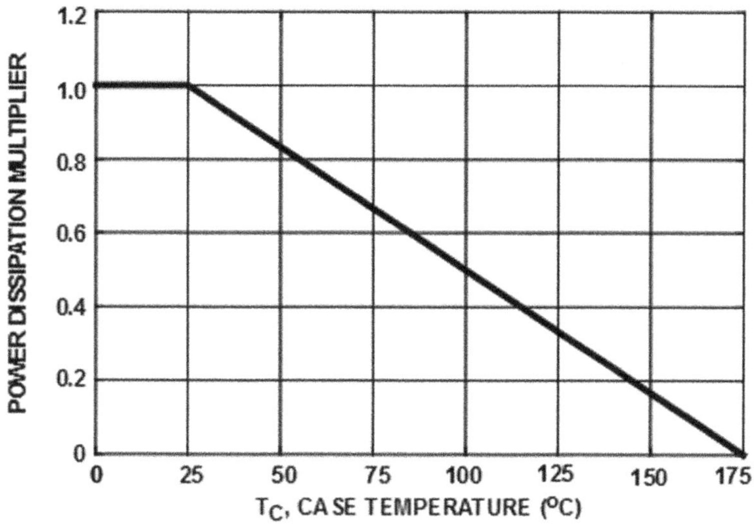

Figure 3.12. *Power derating of a power MOSFET*

3.2.15. *Integrated circuits*

The main stresses are:

– junction temperature;

– supply voltage.

Generally, there are no derating rules imposed by the manufacturers.

3.3. Reference-based approach

There are no standards defining the maximum load rate per family of components; however, methodology guides do exist. Load rates often differ from one guide to another, and sometimes the stress parameters of a family of components are not all identical. Moreover, most of the time, these derating rules are not justified and therefore not readily applicable to a given application. Our first approach was to list all the derating values obtained and to thus draw a common basis, as shown by the following table.

Family of components	Physical contribution	Derating rules
Resistors	Voltage	80%
	Power	60%
Capacitors	Voltage	60%
Connectors	Current	60%
Fuses	Current	50%
Inductances	Current	60%
	Hotspot temperature	60%
Transformers	Current	60%
	Hotspot temperature	60%
	Voltage	60%
Diodes	Reverse voltage	75%
	Junction temperature	70%
Optocouplers	Voltage	75%
	Reverse voltage	75%
	Junction temperature	70%
Bipolar transistors	Voltage Vce	75%
	Voltage Vbe	75%
	Junction temperature	70%
MOSFET transistors	Voltage Vds	75%
	Voltage Vgs	75%
	Junction temperature	70%
Analog circuits	Supply voltage	75%
	Junction temperature	80%
Logic circuits	Supply voltage	75%
	Junction temperature	80%

Table 3.3. *Summary of derating rules according to the literature*

NOTE.– The rates in the table are an average of the rates issued from the following guides:

– MIL-HDBK-338B (1988);

– NAV-SEA (1991);

– MIL-STD-975 NASA (2003).

3.4. Creation of derating rules

The derating rules proposed in this book were established in order to provide a justification for the rules generally used in the design of electronic circuits. This analysis relies on the physics of the failure of components.

It therefore seems appropriate to take into account the reliability level of the component, depending on the physical contribution involved in these derating rules. More specifically, the focus will be on the sensitivity of component reliability with respect to the empirical acceleration law modeling the influence of the considered physical contribution.

As already noted, the derating rate is defined as the ratio of the value applied to the component and the maximum value guaranteed by the manufacturer of the component. It therefore ranges between 0 and 100%. It therefore seems relevant to consider that the maximum acceptable derating DR_Max is applicable to the family of components most sensitive to the physical contribution being considered.

Moreover, it is appropriate to take a minimal derating rate DR_Min to avoid needless overdesign. This derating rate will be assigned to the family of components least sensitive to the applied physical contribution. The value of this margin depends on the type of application (consumer products, aeronautics, automotive and aerospace industries, etc.).

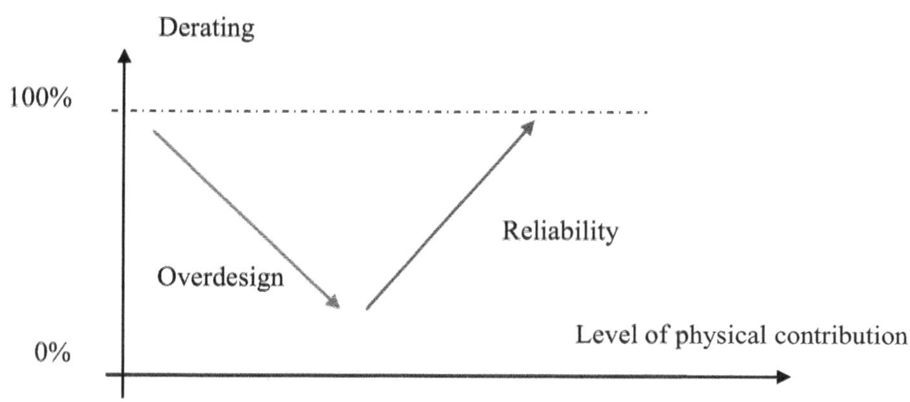

Figure 3.13. *Compromise between reliability and overdesign. For a color version of this figure, see www.iste.co.uk/bayle/maturity1.zip*

The value of these derating rates is the result of a compromise between "overdesign" and a good level of reliability, as illustrated in the figure above.

This qualitative analysis must be followed by a quantitative approach to building these rules. This involves using the sensitivity of a function of a given parameter as the ratio between the relative variation of the function and the relative variation of the considered variable. The sensitivity of the function "f" with respect to the variable x is given by:

$$S_f^x = \frac{\delta f/f}{\delta x/x} \quad [3.5]$$

An equivalent form of the sensitivity of a function is generally expressed as:

$$S_f^p = \frac{p}{f} \cdot \frac{\delta f}{\delta p} \quad [3.6]$$

For the derating, the function "f" is considered as the acceleration factor of the underlying physical law governing the failure. As for the parameter "xp", the characteristic parameter of this law is considered (e.g. the activation energy for the temperature). The following derating rule "DR" is then proposed:

$$DR = a.\left(S_{AF}^p\right)^b \quad [3.7]$$

where "a" and "b" are two parameters to be calculated. The problem is that the parameter "p" can take various values depending on the type of component and therefore on the failure mechanism being considered. Therefore, two extreme cases must be considered, namely the minimum value p_min and the maximum value p_max. The previous qualitative rules lead to the following product of equations:

$$a.\left(S_{AF}^{p\_max}\right)^b = \mathrm{DR\_Max} \quad [3.8a]$$

$$a.\left(S_{AF}^{p\_min}\right)^b = \mathrm{DR\_Min} \quad [3.8b]$$

The values of DR_Max and DR_Min correspond, respectively, to the minimal and maximum values of the derating rules. The method proposed here is first illustrated in Figure 3.14.

A compromise between the robustness margin and overdesign can lead to the following rules:

– DR_Max = 90%;

– DR_Min = 50%.

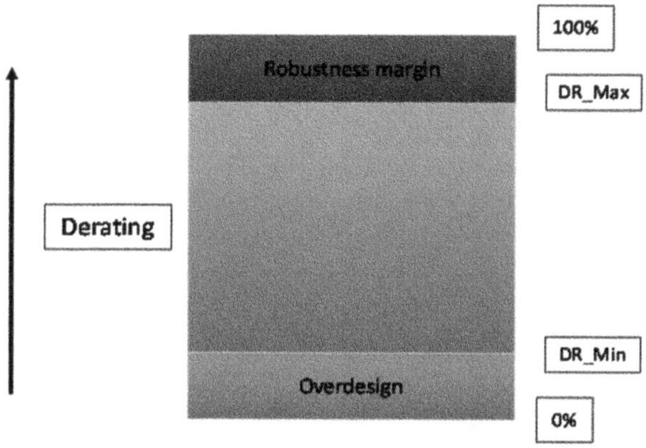

Figure 3.14. *Illustration of derating rules. For a color version of this figure, see www.iste.co.uk/bayle/maturity1.zip*

The next step is then to determine the values of the parameters DR_Max and DR_Min. These values depend on the field of application and should therefore be estimated by those responsible for derating studies in the company concerned. For example, the MIL-HDBK-338B handbook (MIL 1998) provides similar information in the following table.

PART TYPE	DERATING PARAMETER	DERATING LEVEL I (Space)	 II (Airborne)	 III (Ground)
TRANSISTORS				
• Thyristors	On-State Current (I_t – % Rated)	50%	70%	70%
(SCR/TRIAC)	Off-State Voltage (V_{DM} – % Rated)	70%	70%	70%
	Max T (°C)	95°	105°	125°
• Field Effect	Power Dissipation (% Rated)	50%	60%	70%
	Breakdown Voltage (% Rated)	60%	70%	70%
	Max T (°C)	95°	105º	125°
• Bipolar	Power Dissipation (% Rated)	50%	60%	70%
	Breakdown Voltage (% Rated)	60%	70%	70%
	Max T (°C)	95°	105°	125°

For certain applications, such as those in the aerospace sector, where the stakes of reliability, safety, etc. are very high, the following rules can be considered:

– DR_Max = 80%;

– DR_Min = 30%.

The next step is to calculate parameters a and b satisfying the previous product of equations, depending on the physical law of the failure. The result is then:

$$b = \frac{\ln\left(\frac{DR\_Max}{DR\_Min}\right)}{\ln\left(\frac{S_{AF}^{p\_max}}{S_{AF}^{p\_min}}\right)} \quad [3.9]$$

$$a = \frac{DR\_Min}{\left(S_{AF}^{p\_min}\right)^{b}} \quad [3.10]$$

PROOF.–

The previous product of equations leads to:

$$a.\left(S_{AF}^{p\_max}\right)^{b} = DR\_Max$$

$$a.\left(S_{AF}^{p\_min}\right)^{b} = DR\_Min$$

The ratio of the two equations term by term leads to eliminating parameter a. Parameter b can then be calculated as follows:

$$\left(\frac{S_{AF}^{p\_max}}{S_{AF}^{p\_min}}\right)^{b} = \frac{DR\_Max}{DR\_Min}$$

The natural logarithm of the previous equation gives:

$$b.\ln\left(\frac{S_{AF}^{p\_max}}{S_{AF}^{p\_min}}\right) = \ln\left(\frac{DR\_Max}{DR\_Min}\right)$$

or finally

$$b = \frac{\ln\left(\frac{DR\_Max}{DR\_Min}\right)}{\ln\left(\frac{S_{AF}^{p\_max}}{S_{AF}^{p\_min}}\right)}$$

Inserting the value of b in the first equation, the value of parameter “a” is obtained.

$$a.\left(S_{AF}^{p\_min}\right)^{b} = DR\_Min$$

or finally

$$a = \frac{\mathrm{DR\_Min}}{\left(S_{AF}^{p\_min}\right)^{\mathrm{b}}}$$

End

The derating rules are then written for constant temperature as follows:

$$DR(p) = \mathrm{DR\_Min.}\left(\frac{S_{AF}^{p}}{S_{AF}^{p\_min}}\right)^{\mathrm{b}} \quad [3.11]$$

PROOF.–

As already noted: $DR = \mathrm{a.}\left(S_{AF}^{p\_min}\right)^{\mathrm{b}}$ therefore: $DR = \frac{\mathrm{DR\_Min}}{\left(S_{AF}^{p\_min}\right)^{\mathrm{b}}}\cdot\left(S_{AF}^{p}\right)^{\mathrm{b}}$

Hence:

$$DR(p) = \mathrm{DR\_Min.}\left(\frac{S_{AF}^{p}}{S_{AF}^{p\_min}}\right)^{\mathrm{b}}$$

End

3.4.1. *Rules for constant temperature*

The failure mechanisms accelerated by temperature are generally governed by the Arrhenius law (Arrhenius 1889) whose acceleration factor is given by:

$$AF(\theta) = \exp\left(\frac{\mathrm{Ea}}{\mathrm{Kb}}\cdot\left(\frac{1}{\theta\mathrm{o}} - \frac{1}{\theta}\right)\right) \quad [3.12]$$

NOTE.– This is an acceleration with respect to the reference temperature θo.

– Here, T designates the junction temperature θj for the active components and the hotspot temperature Th for the passive components.

Let us then study the sensitivity of this acceleration factor with respect to the activation energy Ea. This can be written as:

$$S_{AF}^{Ea} = \frac{\text{Ea}}{\text{AF}}.\frac{\delta\text{AF}}{\delta\text{Ea}} = \frac{\text{Ea}}{\text{AF}}.\exp\left(\frac{\text{Ea}}{\text{Kb}}.\left(\frac{1}{\theta\text{o}} - \frac{1}{\theta}\right)\right).\left(\frac{1}{\text{Kb}}.\left(\frac{1}{\theta\text{o}} - \frac{1}{\theta}\right)\right)$$

or

$$S_{AF}^{Ea} = \frac{\text{Ea}}{\text{Kb}}.\left(\frac{1}{\theta\text{o}} - \frac{1}{\theta}\right)$$

Consequently, parameters “a” and “b” are written as:

$$\text{b} = \frac{\ln\left(\frac{\text{DR\_Max}}{\text{DR\_Min}}\right)}{\ln\left(\frac{\frac{\text{EaMax}}{\text{Kb}}.\left(\frac{1}{\theta\text{o}}-\frac{1}{\theta}\right)}{\frac{\text{EaMin}}{\text{Kb}}.\left(\frac{1}{\theta\text{o}}-\frac{1}{\theta}\right)}\right)} = \frac{\ln\left(\frac{\text{DR\_Max}}{\text{DR\_Min}}\right)}{\ln\left(\frac{\text{EaMax}}{\text{EaMin}}\right)} \quad [3.13a]$$

$$\text{a} = \frac{\text{DR\_Min}}{\left(\frac{\text{EaMin}}{\text{Kb}}.\left(\frac{1}{\theta\text{o}}-\frac{1}{\theta}\right)\right)^{\text{b}}} \quad [3.13b]$$

The derating rules can then be written for the constant temperature as:

$$DR(Ea) = \text{DR\_Min}.\left(\frac{\text{Ea}}{\text{Ea\_Min}}\right)^{\frac{\ln\left(\frac{\text{DR\_Max}}{\text{DR\_Min}}\right)}{\ln\left(\frac{\text{EaMax}}{\text{EaMin}}\right)}} \quad [3.14]$$

PROOF.–

It is known that: $\text{DR(p)} = \text{DR\_Min}.\left(\frac{\text{S}_{\text{AF}}^{\text{p}}}{\text{S}_{\text{AF}}^{\text{p\_min}}}\right)^{\text{b}}$ and that $\text{S}_{\text{AF}}^{\text{Ea}} = \frac{\text{Ea}}{\text{Kb}}.\left(\frac{1}{\theta\text{o}} - \frac{1}{\theta}\right)$

Combining these two equations leads to:

$$DR(Ea) = \text{DR\_Min}.\left(\frac{\frac{\text{Ea}}{\text{Kb}}.\left(\frac{1}{\theta\text{ref}} - \frac{1}{\theta}\right)}{\frac{\text{Ea\_Min}}{\text{Kb}}.\left(\frac{1}{\theta\text{o}} - \frac{1}{\theta}\right)}\right)^{\text{b}}$$

or finally:

$$DR(Ea) = \text{DR\_Min}.\left(\frac{\text{Ea}}{\text{Ea\_Min}}\right)^{\frac{\ln\left(\frac{\text{DR\_Max}}{\text{DR\_Min}}\right)}{\ln\left(\frac{\text{EaMax}}{\text{EaMin}}\right)}}$$

End

Based on the FIDES guide for predictive reliability (FIDES 2009), the following table presents activation energy as a function of component type.

Types of components	Ea (eV)
Ceramic capacitors	0.1
Electrolytic capacitors	0.4
Tantalum capacitors	0.15
Plastic capacitors	0.25
CMS resistors	0.15
Fuses	0.15
Magnetic components	0.15
Diodes	0.7
Transistors	0.7
Integrated circuits	0.7
LED	0.6
Optocouplers	0.4
ASIC	0.7
Relay	0.25
Interrupters and switches	0.25
Connectors	0.1
Hybrid	0.7
RF and HF Circuits	1.5
LCL STN screen	0.6
LCD TFT screen	0.5
Lithium battery	0.4

Table 3.4. *Activation energy for various types of components*

Therefore, Ea_Min = 0.1 eV and Ea_Max = 0.7eV. Using the proposed derating rules, the following derating rates are obtained.

Types of components	Derating
Ceramic capacitors	50%
Connectors	50%
Tantalum capacitors	54.6%
CMS resistors	54.6%
Fuses	54.6%
Magnetic components	54.6%
Plastic capacitors	61%
Relay	61%
Interrupters and switches	61%
Electrolytic capacitors	67.6%
Optocouplers	67.6%
Lithium battery	67.6%
LCD TFT screen	70.9%
LED	73.8%
LCL STN screen	73.8%
Diodes	76.3%
Transistors	76.3%
Integrated circuits	76.3%
ASIC	76.3%
Hybrid	76.3%
RF and HF circuits	90.0%

Table 3.5. *Temperature derating rate*

This can be illustrated by Figure 3.15.

To illustrate the method, the acceleration factor can be calculated with respect to a maximum temperature of 125°C, and to the derated temperature, as illustrated in Figure 3.16.

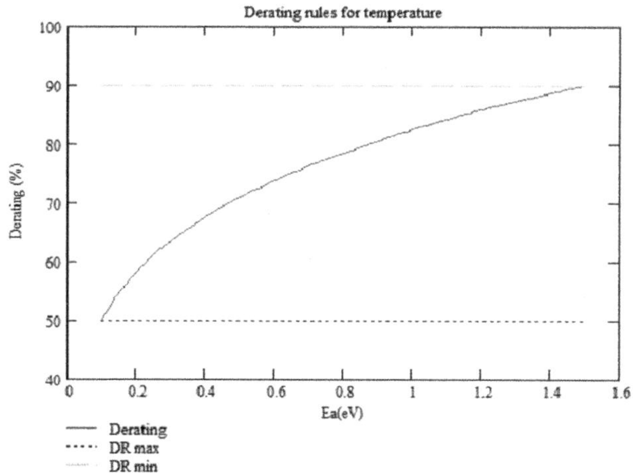

Figure 3.15. *Derating rate depending on activation energy. For a color version of this figure, see www.iste.co.uk/bayle/maturity1.zip*

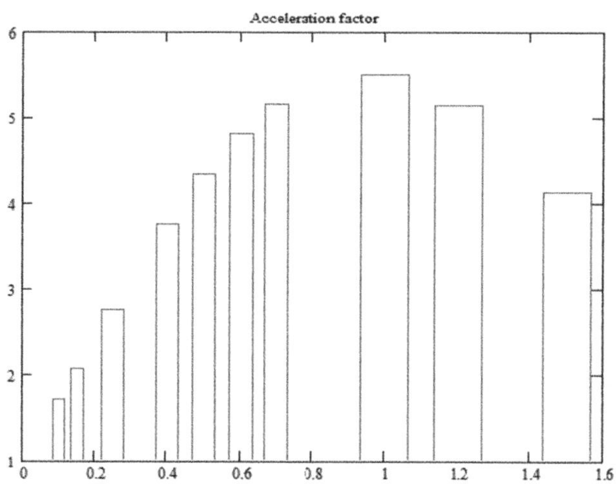

Figure 3.16. *Acceleration factor depending on the derated temperature. For a color version of this figure, see www.iste.co.uk/bayle/maturity1.zip*

It can be noted that the acceleration factors are relatively homogeneous.

3.4.2. *Rule for voltage*

Failure mechanisms accelerated by voltage are generally governed by a power law whose acceleration factor is given by:

$$AF(V) = \left(\frac{\mathrm{V}}{\mathrm{Vref}}\right)^{\epsilon} \qquad [3.15]$$

Let us study the sensitivity of this acceleration factor with respect to parameter ϵ. The following relation can be written:

$$S_{AF}^{\epsilon}(V) = \frac{\epsilon}{\mathrm{AF}} \cdot \frac{\delta \mathrm{AF}}{\delta \epsilon} = \frac{\epsilon}{\mathrm{AF}} \cdot \left(\frac{\mathrm{V}}{\mathrm{Vref}}\right)^{\epsilon} . \ln\left(\frac{\mathrm{V}}{\mathrm{Vref}}\right)$$

or

$$S_{AF}^{\epsilon}(V) = \frac{\epsilon}{\left(\frac{\mathrm{V}}{\mathrm{Vref}}\right)^{\epsilon}} \cdot \left(\frac{\mathrm{V}}{\mathrm{Vref}}\right)^{\epsilon} . \ln\left(\frac{\mathrm{V}}{\mathrm{Vref}}\right) = \epsilon . \ln\left(\frac{\mathrm{V}}{\mathrm{Vref}}\right)$$

This leads to:

$$DR(\epsilon) = \mathrm{DR\_Min.}\left(\frac{\epsilon}{\epsilon\mathrm{\_Max}}\right)^{\mathrm{b}} \qquad [3.16]$$

PROOF.–

It is known that: $DR(\epsilon) = \mathrm{DR\_Min.}\left(\frac{S_{AF}^{\epsilon}}{S_{AF}^{\epsilon\_Max}}\right)^{\mathrm{b}}$ and that $S_{AF}^{\epsilon}(\mathrm{T}) = \epsilon . \ln\left(\frac{\mathrm{V}}{\mathrm{Vref}}\right)$

Combining these two equations leads to:

$$DR(\epsilon) = \mathrm{DR\_Min.}\left(\frac{\epsilon . \ln\left(\frac{\mathrm{V}}{\mathrm{Vref}}\right)}{\epsilon\mathrm{\_Max} . \ln\left(\frac{\mathrm{V}}{\mathrm{Vref}}\right)}\right)^{\mathrm{b}}$$

or

$$DR(\varepsilon) = \mathrm{DR\_Min.}\left(\frac{\epsilon}{\epsilon\mathrm{\_Max}}\right)^{\frac{\ln\left(\frac{\mathrm{DR\_Max}}{\mathrm{DR\_Min}}\right)}{\ln\left(\frac{\epsilon\mathrm{\_Max}}{\epsilon\mathrm{\_Min}}\right)}}$$

End

3.5. Summary

The proposed method for the estimation of derating rules relies on the physical laws of failure. Unfortunately, the parameters of a component that are guaranteed by the manufacturer cannot all be modeled by the laws of physics. Therefore, in order to have a summary of the various rules proposed (by the manufacturer, by standards and by the proposed method), only the parameters meeting this constraint are listed in the following table.

Family of components	Physical contribution	Derating references	Proposed derating
Chip resistors	Hotspot temperature	80%	54.6%
Ceramic capacitors	Hotspot temperature	60%	50%
	Direct voltage	60%	71.7%
Electrolytic capacitors	Hotspot temperature	60%	67.6%
	Direct voltage	90%	71.7%
Tantalum capacitors	Hotspot temperature	60%	54.6%
	Direct voltage	60%	71.7%
Transformers	Hotspot temperature	60%	54.6%
Inductances	Hotspot temperature	60%	54.6%
Connectors	Hotspot temperature	Tmax – 25%	50%
Fuses	Hotspot temperature	50%	54.6%
Diodes	Junction temperature	70%	76.3%
Optocouplers	Junction temperature	70%	67.6%
Bipolar transistors	Junction temperature	70%	76.3%
Analog circuits	Junction temperature	80%	76.3%
Logic circuits	Junction temperature	80%	76.3%
RF and HF circuits	Junction temperature	80%	90%

Table 3.6. *Summary of derating rules*

4

Components with Limited Service Life

As already noted, a product is considered mature provided no aging mechanisms are observed when it is put into operation. However, all components have several potential aging mechanisms. Let us take the example of complex integrated circuits; there are failure mechanisms that may be generated at three different locations:

– the chip:

  - hot carrier injection (HCI),

  - negative bias thermal instabilities (NBTI),

  - positive bias thermal instabilities (PBTI),

  - electromigration (EM);

– the interface between chip and package:

  - bonding lift-off,

  - bonding breakage,

  - degradation of the braze between chip and substrate;

– the component interface between the printed circuit and the pins of the component:

  - breakage of the brazed joints.

However, experience shows that if the component is correctly designed, these aging mechanisms are never observed, except for certain categories of components which have limited service life under certain conditions.

Hence, using the bill of materials of the product under consideration, it is absolutely essential to check whether or not these mechanisms will be observed under the operational conditions to which the component will be subjected.

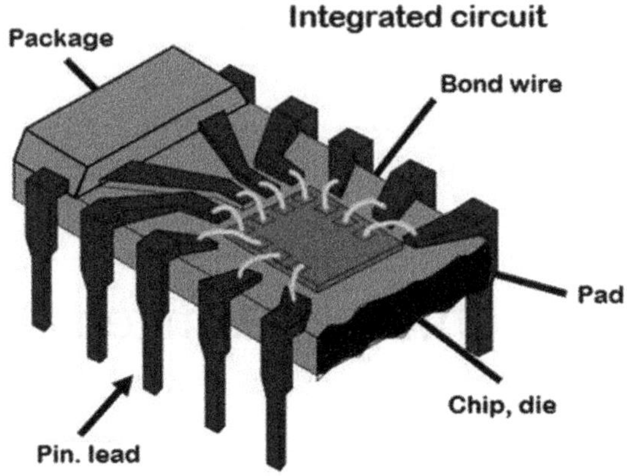

Figure 4.1. *Overview diagram of an integrated circuit. For a color version of this figure, see www.iste.co.uk/bayle/maturity1.zip*

Certain predictive reliability methods (RDF 2000; FIDES 2009) include a brief warning as to whether this is the case or not. The service life of components in storage is beyond the scope of this chapter.

The following figure illustrates this subject in the case of the FIDES guide.

Light-emitting diode (LED)

General model associated with the family ① Warning: limited service life

$\lambda = \lambda_{Physical} \times \Pi_{PM} \times \Pi_{Process}$ with:

$$\lambda_{Physical} = \sum_{i}^{Phases} \left(\frac{t_{annual}}{8760}\right)_i \times \begin{pmatrix} \lambda_{0TH} \times \Pi_{Thermal} \\ + \lambda_{0TCyCase} \times \Pi_{TCyCase} \\ + \lambda_{0TCyBrazed\,joints} \times \Pi_{TCyBrazed\,joints} \\ + \lambda_{0RH} \times \Pi_{RH} \\ + \lambda_{0Mech} \times \Pi_{Mech} \end{pmatrix}_i \times (\Pi_{Induced})_i$$

Figure 4.2. *Example of components with limited service life according to FIDES guide*

On the other hand, for some categories of components, manufacturers specify certain constraints on one or several key parameters of the component. The following table proposes a summary of the components with limited service life.

Types of components
Optocouplers
LED
Wet electrolytic capacitor
Electromechanical relays
Interrupters/switches
Batteries
Fans
Connectors
Power transistors
Flash memory
Potentiometers
Quartz oscillator
Voltage regulators

Table 4.1. *List of components with limited service life*

4.1. RDF 2000 guide

4.1.1. *Power transistor*

According to the RDF 2000 guide, cyclic power transistors have a limited service life. The number of cycles that should not be exceeded is:

$$N(\Delta T) = \exp(-0.05.\Delta\mathrm{T}) \quad [4.1]$$

This yields the following figure.

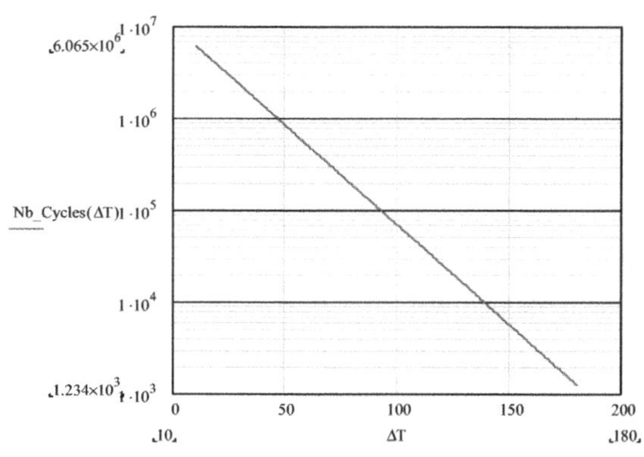

Figure 4.3. *Number of cycles depending on the thermal amplitude of the power transistor (RDF 2000). For a color version of this figure, see www.iste.co.uk/bayle/maturity1.zip*

4.1.2. *Photocouplers*

According to the RDF 2000 guide, photocouplers have a limited service life. The number of cycles that should not be exceeded is:

$$SL(\theta j) = \exp\left(\frac{4640}{273+\theta j}\right) \quad [4.2]$$

The service life, SL, is expressed in hours and is defined for a 10% proportion of failure. This yields the following figure.

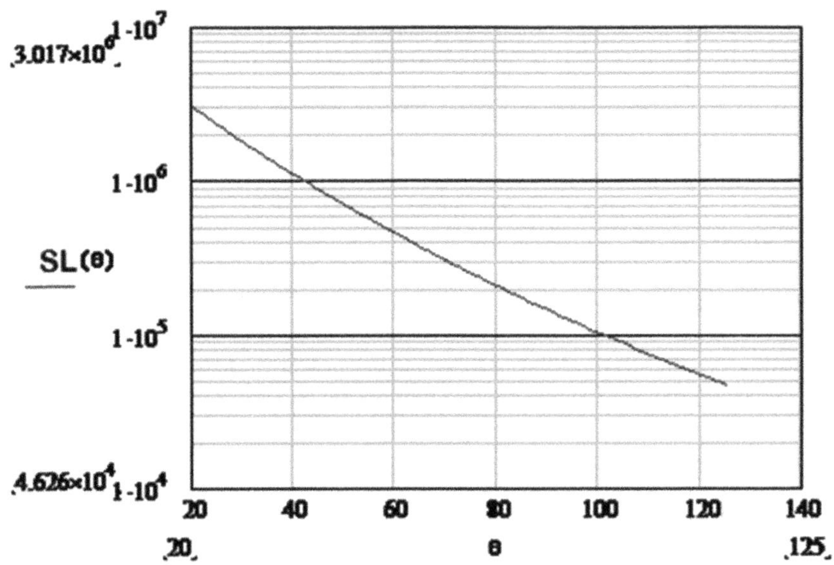

Figure 4.4. *Service life in hours for photocouplers according to the RDF 2000. For a color version of this figure, see www.iste.co.uk/bayle/maturity1.zip*

4.1.3. *Switch or push button*

According to the RDF 2000 guide, a switch or push button has a limited service life, as detailed in the table below.

Service life limitation
The failure rates mentioned here can only be used for a number of operations below the number provided by specification. *Illustration:* • For rocking switches or pushbuttons: 20,000 or 100,000 (sensitive switches) • For rotary-type switches: 20,000 • For keyboards: 500,000 to 1,000,000

Figure 4.5. *Service life of switches or push buttons according to the RDF 2000*

4.1.4. *Connectors*

According to the RDF 2000 guide, connectors have a limited service life, as shown in the table below.

WARNING!
Service life is limited to the number of matings/unmatings mentioned in the specification or the default in the catalogue (several dozen to several hundred).

Figure 4.6. *Service life of connectors according to RDF 2000*

4.2. FIDES 2009 guide

The FIDES guide (2009) indicates the components that are likely to have a limited service life as shown in the following figure.

Interrupters and switches

General model associated with a family: **ⓘ Warning: Limited service life**

Figure 4.7. *Example of potential limited service life of interrupters and switches according to FIDES*

This limited service life is not taken into account when calculating the failure rate of the component. According to FIDES, designers are responsible for making sure this is not the case in the considered design:

– Components with a limited service life have an "aging" failure rate added to the component failure rate expression.

$$\lambda = \lambda_{\text{catastrophic}} + \lambda_{\text{aging}} \qquad [4.3]$$

This "aging" failure rate takes into account the effects of preventive maintenance (PM) or corrective maintenance (CM).

4.2.1. *Fans*

The FIDES guide (2009) provides service life data for fans. These are defined by the parameter L10, often used in mechanics, which is the time at which 10% of failures were observed. The following table illustrates these data.

		Merit factor (Wh/Kg)	NbCD: No of Charge/ Discharge cycles	SL in calendar hours
Lithium	Li-ion Polymer	-	300	20 000
	Li-ion (1)	≤ 100	300	20 000
		100 < ≤ 200	500	
		> 200	1 000	
	Li-metal Phosphate	-	2 000	35 000
	Nano Titanium	-	10 000	150 000
Nickel	NiMH Nickel metal hydride (2)	≤ 50	300	NA
		> 50	1 000	
	Ni-Zn Zinc	-	1 000	
	Ni-Cd Cadmium	-	2 000	

(1) and (2): If the merit factor is not known, consider NbCD = 500

$$DDV_1 = \frac{NbCD}{NbCD_{Yearly}} \times 8760$$ (8760, number of hours per year for a yearly life profile)

with $NbCD_{Yearly}$: Number of charge/discharge cycles per year

Table 4.2. *Service life of batteries according to the FIDES guide*

Moreover, this guide proposes a 0.4 eV activation energy.

4.2.2. *Batteries*

The table below presents an example of the estimated service life for batteries.

$$\lambda_{Aging} = 0.105 \cdot 10^{9} \times \frac{\min(MPP, MC)^{\beta - 1}}{\min(SL_1, SL_2)^{\beta}}$$

with:

- SL_1, SL_2 calendar service lives
  The batteries have a limited service life (SL_1) related to the number of charge/discharge cycles. In addition, and independently of the number of charge/discharge cycles, lithium batteries have a service life (SL_2) that is limited in time. For the estimation of service lives, the failure criterion is generally the capacity of the battery below a specified threshold (generally between 60% and 80% of the initial capacity).

In the absence of manufacturer data, consider:

		Merit factor (Wh/Kg)	NbCD: No of Charge/ Discharge cycles	SL in calendar hours
Lithium	Li-ion Polymer	-	300	20 000
	Li-ion (1)	≤ 100	300	20 000
		100 < ≤ 200	500	
		> 200	1 000	
	Li-metal Phosphate	-	2 000	35 000
	Nano Titanium	-	10 000	150 000
Nickel	NiMH Metal Hydride (2)	≤ 50	300	NA
		> 50	1 000	
	Ni-Zn Zinc	-	1 000	
	Ni-Cd Cadmium	-	2 000	

(1) and (2): If the merit factor is not known, consider NbCD = 500

$$SL_1 = \frac{NbCD}{NbCD_{Yearly}} \times 8760$$ (8760, number of hours per year for a yearly life profile)

with $NbCD_{Yearly}$: Number of charge/discharge cycles per year

		Merit factor (Wh/Kg)	NbCD: No. of Charge/Discharge cycles	SL in calendar hours
Lithium	Li-ion Polymer	-	300	20 000
	Li-ion (1)	≤ 100	300	20 000
		100 < ≤ 200	500	
		> 200	1 000	
	Li-metal Phosphate	-	2 000	35 000
	Nano Titanium	-	10 000	150 000
Nickel	NiMH Metal Hydride (2)	≤ 50	300	NA
		> 50	1 000	
	Ni-Zn Zinc	-	1 000	
	Ni-Cd Cadmium	-	2 000	

Figure 4.8. *Service life of batteries according to FIDES*

4.3. Manufacturer's data

Depending on the family of components, the manufacturers specify the service life of certain components.

4.3.1. *Wet electrolytic capacitor*

These capacitors are considered to be components with limited service life, most notably because the value of their key characteristic, that is, capacitance (or series resistance, depending on the application), decreases over time and with temperature. This phenomenon occurs because, with time and temperature, the electrolyte inside the capacitor tends to evaporate. The service life during storage is beyond the scope of this section.

This is generally reflected in the datasheet of the component by the concept of endurance, as illustrated below (PAN 2019).

	+105 ℃ ± 2 ℃, 10000 h, apply the rated ripple current without exceeding the rated voltage.	
Endurance	Capacitance change	Within ±30% of the initial value
	Dissipation factor (tan δ)	≦ 200 % of the initial limit
	ESR	≦ 200 % of the initial limit
	DC leakage current	Within the initial limit

Table 4.3. *Endurance parameter of a wet electrolytic capacitor*

Hence, there are four failure criteria:

– a 30% variation of capacitance, generally a decrease;

– a 200% variation of the tangent δ;

– a 200% variation of the series resistance;

– a leakage current in accordance with the specification.

If at least one of these criteria is not met, a failure occurs. Hence, the service life at 105°C, under specified current and voltage conditions, is 10,000 hours. The problem is that this data is not sufficient to know the service life for another temperature; this is because of the absence of the acceleration law.

It is possible to rely on a predictive reliability model such as FIDES (2009) in order to get this information. The considered activation energy is Ea = 0.4 V.

Assume that for our application, the temperature is 45°C. The corresponding service life can then be written as:

$$SL@45°C = \text{SL@105°C. AF}$$

or numerically

$$SL@45°C = 10\,000.\exp\left(\frac{\text{Ea}}{\text{Kb}}.\left(\frac{1}{273+40}-\frac{1}{273+105}\right)\right) \sim 101\,457 \text{ hours}$$

A further solution is to use the following empirical law:

$$SL(T) = \text{DDVo}.\,2^{\frac{\text{To}-\text{T}}{10}}$$

Resuming the previous example, this yields:

$$SL(45) = 10\,000.\,2^{\frac{105-45}{10}} = 640\,000 \text{ hours}$$

It can be seen that the two models yield very different sum results. The following figure compares them in terms of acceleration, with respect to the maximal temperature (here, 105°C).

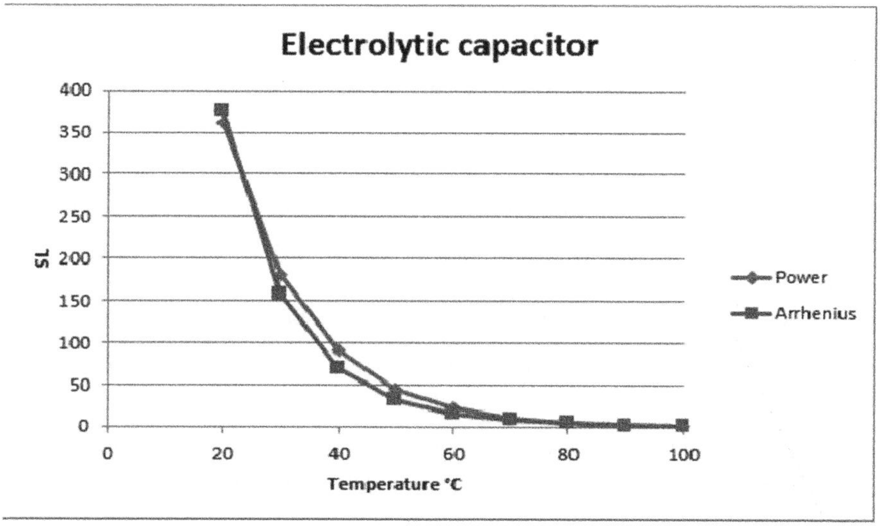

Figure 4.9. *Comparison of the models of service life of an electrolytic capacitor. For a color version of this figure, see www.iste.co.uk/bayle/maturity1.zip*

An attempt was then made to find an equivalence between the two models. This can be established by adjusting the activation energy of the Arrhenius model. Minimizing the sum of squared errors between the two models yields an approximately equivalent activation energy of 0.665 eV, which is quite different from that of 0.4 eV. The following figure illustrates the quality of the equivalence.

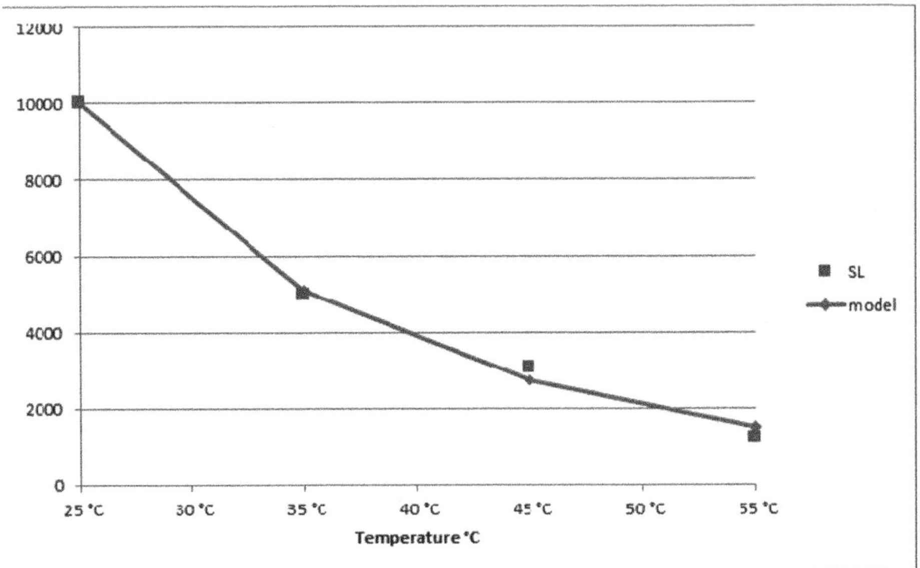

Figure 4.10. *Equivalence between the two models of service life of an electrolytic capacitor. For a color version of this figure, see www.iste.co.uk/bayle/maturity1.zip*

Finally, in the absence of additional data that could be used to choose one model over another, three cases may arise for the studied application:

– for the two models, the capacitor presents no service life problems. The analysis is therefore completed and the capacitor is suitable for this application;

– for the two models, the capacitor presents service life problems. Therefore, the component should be changed or its temperature reduced;

– the power model accepts the component, while the Arrhenius model rejects it. As a precautionary measure, changing the component or reducing its temperature is recommended.

4.3.2. *Connectors*

Connectors are components that are generally considered to have a limited service life as a result of their limited number of "insertions/disinsertions". This parameter, durability (max. mating cycles) is mentioned in the connector datasheet (200 for this type of connector) as illustrated below.

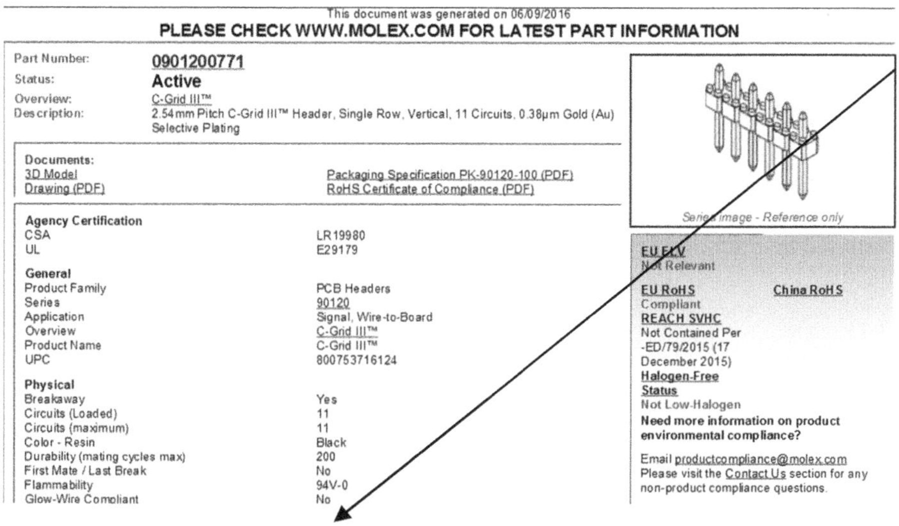

This document was generated on 06/09/2016

PLEASE CHECK WWW.MOLEX.COM FOR LATEST PART INFORMATION

Part Number:	0901200771
Status:	**Active**
Overview:	C-Grid III™
Description:	2.54mm Pitch C-Grid III™ Header, Single Row, Vertical, 11 Circuits, 0.38µm Gold (Au) Selective Plating

Documents:
3D Model
Drawing (PDF)
Packaging Specification PK-90120-100 (PDF)
RoHS Certificate of Compliance (PDF)

Agency Certification	
CSA	LR19980
UL	E29179
General	
Product Family	PCB Headers
Series	90120
Application	Signal, Wire-to-Board
Overview	C-Grid III™
Product Name	C-Grid III™
UPC	800753716124
Physical	
Breakaway	Yes
Circuits (Loaded)	11
Circuits (maximum)	11
Color - Resin	Black
Durability (mating cycles max)	200
First Mate / Last Break	No
Flammability	94V-0
Glow-Wire Compliant	No

Series image - Reference only

EU ELV
Not Relevant
EU RoHS
Compliant
China RoHS
REACH SVHC
Not Contained Per -ED/79/2015 (17 December 2015)
Halogen-Free Status
Not Low-Halogen
Need more information on product environmental compliance?
Email productcompliance@molex.com
Please visit the Contact Us section for any non-product compliance questions.

Figure 4.11. *Example of the maximal number of insertions/disinsertions for a connector. For a color version of this figure, see www.iste.co.uk/bayle/maturity1.zip*

In practice, it is not always easy to estimate the number of insertions/ disinsertions. In fact, this may occur at a number of stages of the product lifecycle, such as:

– burn-in;

– acceptance test before delivery to the system manufacturer;

– integration stage at the system manufacturer;

– product operation during maintenance.

A certain margin is therefore recommended between the evaluation conducted and the maximal capacity of the connector.

4.3.3. *Relays*

Relays are generally considered to be components with a limited service life because of the limited number of operations (on/off). This number depends on several parameters such as:

– load type (resistive, capacitive or inductive);

– contact type (gold, silver, etc.);

– current through the component when it is off;

– etc.

The information provided on the service life of relays does not always take all of these parameters into account, but it is available for various types of loads, as illustrated below.

Minimal service life	Load type
100,000 cycles	Resistive
20,000 cycles	inductive
400,000 cycles	Below 25% of the rated resistive load

Figure 4.12. *Example of relay service life*

Similar to relays, this information is not always easy to estimate in practice. When the relay is activated for a well-known operation, this is still possible. When it acts as a protective, backup device, the events that may activate it are rather random and therefore difficult to record.

4.3.4. *Optocouplers*

Data on the service life and reliability of an optocoupler are not easily accessible. In contrast to chemical capacitors, for example, they are never mentioned in the technical specifications of the optocouplers. Some factors degrade the service life of an optocoupler significantly. To extend service life to span decades (no need for it to exceed the service life of the product!), the following criteria must be met:

– the LED of the optocoupler should be supplied with a low current;

– the voltage between LED and phototransistor (voltage across the isolation barrier) should not be too high;

– very hot environments should be avoided (temperature is the enemy).

It is therefore important to size the circuit and limit the LED current as much as possible. The lower the current through the collector, the better.

It is also important to size the circuit with a good margin on CTR (Current Transfer Ratio) as the optocoupler will progressively diminish throughout its operational life. The definition is as follows (Toshiba 2018).

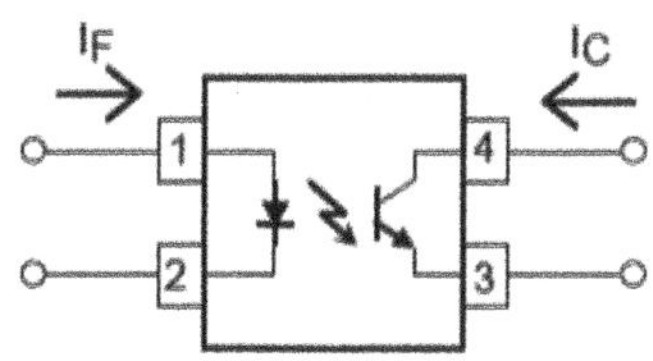

$$CTR = 100 \times I_C / I_F \ [\%]$$

Figure 4.13. *Illustration of optocoupler CTR*

Moreover, there is often wide variability in the CTR range from one optocoupler to another, as illustrated in Figure 4.13 (Toshiba 2018).

Figure 4.15 provides an example of CTR degradation over time (Toshiba 2018).

It shows that:

– on average, CTR decreases over time, indicating an aging mechanism;

– CTR dispersion increases with time.

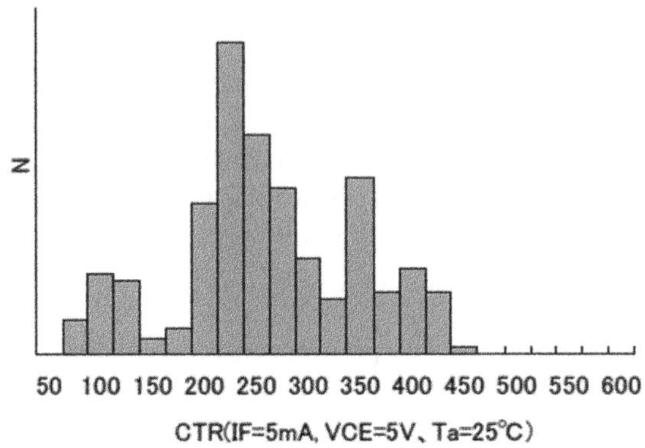

Figure 4.14. *Dispersion of optocoupler CTR*

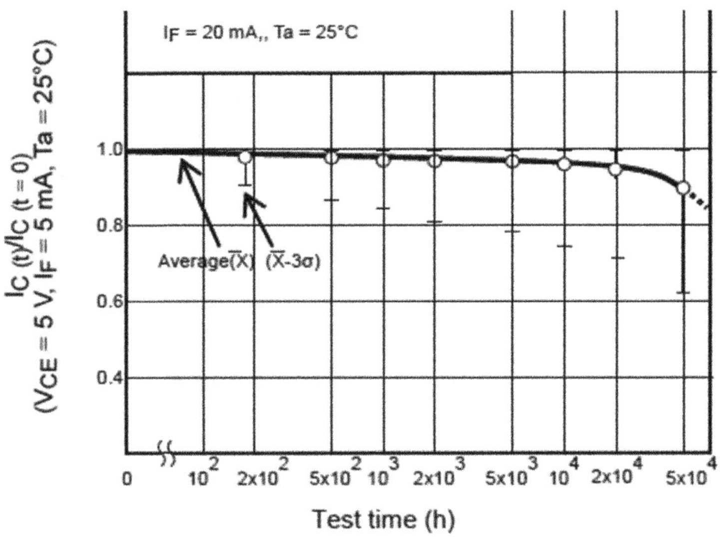

Figure 4.15. *Example of service life data of an optocoupler*

The effect of temperature must also be taken into account. The FIDES guide (2009) proposes an activation energy of 0.4 eV. Another source (AVAGO 2018) indicates an activation energy of 0.43 eV, as illustrated below.

Acceleration Factor Based on Black Model

An Acceleration Factor (AF) based on the Black Model can be used to correlate the actual HTOL stress t taken at elevated temperatures and stress levels in short periods of time, to the expected lifetime accordi application operating conditions of the optocoupler.

$$AF = \left(\frac{J_{acc}}{J_{norm}}\right)^N \exp\left(\frac{E_a}{K}\left\{\frac{1}{T_{norm}} - \frac{1}{T_{acc}}\right\}\right)$$

AF = Acceleration Factor

J_{acc} = Accelerated current density (HTOL stress input current)

J_{norm} = Nominal operation current density (application operating input current at 100% duty cycle)

Ea = Activation energy of 0.43 eV

K = Boltzmann's constant of 8.62×10^{-5} eV/K

T_{norm} = Nominal operating temperature (application ambient operating temperature)

T_{acc} = Accelerated operating temperature (HTOL stress temperature)

N = Model parameter of 2

Table 4.4. *Black model for optocouplers*

EXAMPLE.–

Consider the following example:

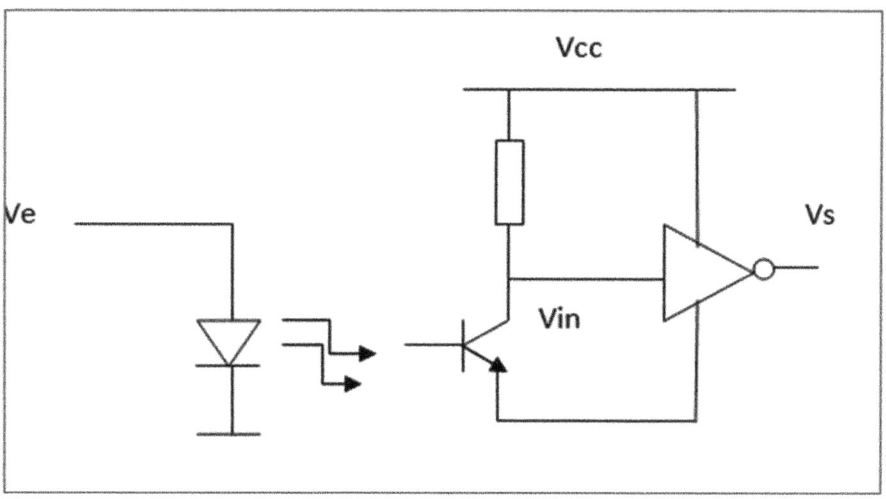

Figure 4.16. *Example of the service life of an optocoupler*

– when Ve = 1, the phototransistor is on, and the voltage Vin = 0 so that Vs = 1;

– when Ve = 0, the phototransistor is blocked, and the voltage Vin = 1 so that Vs = 0;

– the resistance is calculated so that the voltage Vin is below the logic level of the inverter.

Due to time and temperature, the CTR parameter decreases and leads to an increase in voltage Vin. If it decreases too much, the voltage Vin exceeds the logic level 0 of the inverter and the voltage Vs differs from 1. In this case, the proposed diagram no longer fulfills its function and there is a risk of believing that the voltage Ve may change, though that is not the case.

4.3.5. *Batteries*

There are a large number of batteries with different technologies, and it is therefore impossible to provide an exhaustive list. Nevertheless, batteries have a limited service life because of the number of charging/discharging cycles and as a result of temperature.

MDPI (2019) proposes a similar type of table in the following form.

Energy Storage Technology	Specific Energy (Wh/kg)	Specific Power (kW/kg)	Temp. Range (°C)	Cycle life (-)
Mechanical Energy Storage				
Flywheel Energy Storage Systems (FESS)	10–100	>10	−30 °C to +70 °C	limitless
Chemical Energy Storage (only Li-Ion Batteries)				
Lithium Iron Phosphate (LFP)	90–120	4	−20 °C to +60 °C	5000–6000
Lithium Titanate LTO	60–80	1	−30 °C to +75 °C	>15,000
Lithium Nickel Cobalt Aluminum oxide (NCA)	200–300	1	−20 °C to +60 °C	500
Lithium Cobalt Oxide (LCO)	150–200	1	−20 °C to +60 °C	500–1000
Lithium Manganese Oxide (LMO)	100–150	4	−20 °C to +60 °C	300–700
Lithium Nickel Manganese Cobalt oxide (NMC)	150–280	1–4	−20 °C to +55 °C	3000–4000
Electrical Energy Storage				
Double Layer Capacitor (DLC)	5–10	>10	−20 °C to +60 °C	1 million
Hybrid Capacitor	10–20	>10	−20 °C to +60 °C	>50,000

Table 4.5. *Service life of batteries according to MDPI*

As for lithium batteries, frequently used in transportation vehicles, they provide the information as provided in Table 4.6 (MDPI 2019).

This data can be modeled using an Arrhenius law with Ea ~ 0.53 eV.

Temperature	SL
25°C	10 000 hrs
35°C	5 000 hrs
45°C	3 050 hrs
55°C	1 200 hrs

Table 4.6. *Service life of lithium batteries according to MDPI*

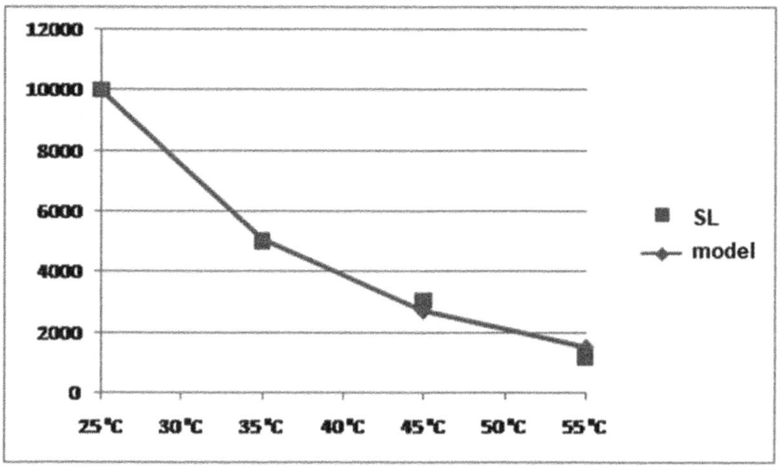

Figure 4.17. *Service life of lithium batteries (MDI). For a color version of this figure, see www.iste.co.uk/bayle/maturity1.zip*

4.3.6. *Fans*

Similar to relays, fans involve various technologies with different service lives, as illustrated in the table below.

Type of fan bearing	Recommendation	SL (25°C)
Sleeve bearing	Operating temperature 25–40°C Take cost into account	25,000 hrs
A ball and a sleeve	Ambient temperature is above 40°C	35,000 hrs
Two ball bearings	Ambient temperature is above 40°C	50,000 hrs
One ball bearing	1 Environmental temperature is above 40°C 2 Suitable with the Slim fan	5,000 hrs

Table 4.7. *Service life of various types of fans (TITAN)*

The service life of fans generally relates to the bearing that allows the rotation of blades, which in turn generates air flow

$$SL = 79200.\Pi_{type}.[3,53 - 0,744.ln(B)].exp\left(-Ea.11604.\left(\frac{1}{313} - \frac{1}{273 +}\right)\right).\left(\frac{V}{3000}\right)^{-m}$$

Type of bearing	Πtype: AC	Πtype: DC	Ea	m
Sealed sleeve bearing	0.5	0.46	0.46	1.6
Single ball bearing (ball/sleeve) Hydrodynamic (hypro)	0.7	0.64	0.4	1.23
Ball bearing (dual)	1	0.92	0.28	0.93
Ceramic bearing	1.4	1.3	0.21	0.57

Table 4.8. *Activation energy of fans according to FIDES*

4.3.7. *Flash memories*

Features

- Organization:
  - Page size:
    x8: 2,112 bytes (2,048 + 64 bytes)
    x16: 1,056 words (1,024 + 32 words)
  - Block size: 64 pages (128K + 4K bytes)
  - Device size: 2Gb: 2,048 blocks; 4Gb: 4,096 blocks; 8Gb: 8,192 blocks
- Read performance:
  - Random read: 25μs
  - Sequential read: 30ns (3V x8 only)
- Write performance:
  - Page program: 300μs (TYP)
  - Block erase: 2ms (TYP)
- Endurance: 100,000 PROGRAM/ERASE cycles
- Data retention: 10 years

Table 4.9. *Service life of a flash memory*

There are two types of flash memory, namely NAND and NOR. A NOR flash memory offers high reliability but low memory density. NAND flash memories have high memory capacities but, depending on how they are used, they may be subject to aging phenomena. Nowadays, these memories are frequently used in telephones, computers, tablets, external hard disks, etc. Their service life is often limited due to the fact that:

– the number of read/write cycles is limited;

– data retention time is not infinite.

Table 4.9 (MIC 2004) illustrates this data.

4.3.8. *Potentiometers*

Potentiometers are used to manually vary resistance value. Due to this manipulation, they are generally considered to be components with potentially limited service life. Two large families can be considered:

– Trimmers, which are generally embedded in a card and whose adjustment is generally done only once. This is due to the fact that the theoretical resistance value is not possible in the list of standard resistance values. An example of a trimmer is given in the figure below.

Figure 4.18. *Trimmer illustration. For a color version of this figure, see www.iste.co.uk/bayle/maturity1.zip*

– Generally, rotary potentiometers (though there are also linear ones, such as the mixer consoles) can be used to vary the resistance value manually. An example is the volume control of a hifi system or a radio receiver. This is why these can be manipulated numerous times.

Figure 4.19. *Illustration of a rotary potentiometer*

The service life of these two types of components is expressed in terms of “mechanical life” or “endurance life”.

It is important to note that these components are being used less and less as they are being replaced by digital potentiometers controlled by an I2C bus, for example, or by an infrared sensor by means of remote control.

4.3.8.1. *Trimmers*

The table below gives an example of service life guaranteed by the manufacturer. It can be seen that the number of manipulations (cycles) is small (200 in this example). As already noted, because they are not handled very much, this number is generally sufficient.

PERFORMANCES						
TESTS	**CONDITIONS**	**MAX. (R)**	**CHANGE PER CECC**		**PER IEC**	**PER MIL**
			V_{AB}/V_{AC}	**41100**		
Vibration	98 m/s², 10 Hz to 500 Hz	1 %	2 %	(PARA 2.3.2)	Test FC (IEC 6-2-6)	Method 204
Electrical endurance	1000 h	3 %	-	(PARA 2.5.16)	-	No equiv.
Soldering	-	-	-	(PARA 2.3.7)	Test TB (IEC 68-2-20)	Method 208
Resistance to heat	-	1 %	-	(PARA 2.3.7)	Test B (IEC 68-2-20A)	Method 210
Damp heat steady state	21 days	3 %	-	(PARA 2.1)	Test C (IEC 68-2-3)	Method 103
Mechanical life	200 cycles	3 %	-	-	Method 2	-
Terminal strength	2.2 lbs. (1 kg)	min.	-	-	-	-

Note

Table 4.10. *Example of limited service life of a trimmer*

4.3.8.2. *Rotary potentiometers*

Service life is specified in terms of mechanical endurance (VIS 2016), as illustrated in the following table.

PERFORMANCE				
TESTS	**CONDITIONS**	**TYPICAL VALUES AND DRIFTS**		
		$\Delta R_T/R_T$ (%)	**$\Delta R_{1\text{-}2}/R_{1\text{-}2}$ (%)**	**OTHER**
Electrical endurance	1000 h at rated power 90'/30' - ambient temp. 70 °C	± 1 %	-	Contact res. variation: < 3 % Rn
Climatic sequence	Phase A dry heat 125 °C Phase B damp heat Phase C cold -55 °C Phase D damp heat 5 cycles	± 0.5 %	± 1 %	-
Damp heat, steady state	56 days 40 °C 93 % HR	± 0.5 %	± 1 %	Insulation resistance: > 10^4 MΩ
Change of temperature	5 cycles -55 °C at +125 °C	± 0.5 %	-	-
Mechanical endurance	25 000 cycles	± 3 %	-	Contact res. variation: < 2 % Rn
Shock	50 g's at 11 ms 3 successive shocks in 3 directions	± 0.1 %	± 0.2 %	-
Vibration	10 Hz to 55 Hz 0.75 mm or 10 g's during 6 h	± 0.1 %	± 0.2 %	-

Table 4.11. *Example of limited service life of a rotary potentiometer*

In this example, the manufacturer guarantees 25,000 manipulations. Similar to the interrupters and switches, it is not always easy to estimate the number of manipulations in the operational life of the component. A useful exercise in this sense is an attempt to estimate the number of times per year the volume of a car radio is manipulated.

4.3.9. *Quartz oscillators*

Strictly speaking, quartz oscillators have no actual malfunction. However, a frequency tolerance above the one specified may have an important operational impact and therefore may be considered a failure.

The table below gives the characteristics in terms of aging (NDK 2003).

■ **Specifications**

Item / Model	NX3225SA			
Standard		Standard		Optional
Nominal Frequency (MHz)	12 to 64	16 to 54	40 to 150	12 to 64
Overtone Order	Fundamental	Fundamental	3rd overtone	Fundamental
Frequency Tolerance (25 ±3 °C)	$\pm15 \times 10^{-6}$	$\pm10 \times 10^{-6}$	$\pm20 \times 10^{-6}$	$\pm10 \times 10^{-6}$
Frequency versus Temperature Characteristics (with reference to +25 °C)	$\pm25 \times 10^{-6}$	$\pm10 \times 10^{-6}$	$\pm25 \times 10^{-6}$	$\pm25 \times 10^{-6}$ (Temp extended case, *1)
Operating Temperature Range (°C)	−40 to +85	−20 to +75	−40 to +85	−40 to +85 *1
Storage Temperature Range		−40 to +85		−40 to +85
Equivalent Series Resistance	Refer to *2	Refer to *3	Refer to *4	Refer to *1
Level of Drive (μW)		10 (Max. 200)		10 (Max. 200)
Load Capacitance (pF)	8	10	Series resonance	6 to 32
Frequency Aging (+25°C)		---		Max. $\pm3 \times 10^{-6}$ / year *1
Specifications Number	STD-CSR-6	STD-CSQ-1	STD-CSR-7	Refer to *5

Table 4.12. *Aging specifications of a quartz oscillator*

The drift in maximal frequency is ± 3 ppm/year at 25°C. The temperature drift, which is 25 ppm throughout the temperature range, should also be taken into account. In the absence of additional data, a worst-case analysis is used to check if the requirement is met. Further details about this analysis are provided in Chapter 8.

4.3.10. *Voltage references*

Similar to quartz oscillators, voltage references can be considered to be a component with limited service life when the voltage drift exceeds that which is specified. The table below gives the characteristics in terms of aging.

Electrical Characteristics (MAX6018_12–1.263V)

(V_{IN} = 1.8V; C_{OUT} = 47nF, I_{OUT} = 0; T_A = T_{MIN} to T_{MAX}, unless otherwise noted. Typical values are at T_A = +25°C.) (Note 1)

PARAMETER	SYMBOL	CONDITIONS	MIN	TYP	MAX	UNITS
OUTPUT						
Output Voltage	V_{OUT}	MAX6018A_12 (0.2%)	1.2605	1.2630	1.2655	V
		MAX6018B_12 (0.4%)	1.2580	1.2630	1.2681	
Output Voltage Temperature Drift	TCV_{OUT}	(Note 2)		16	50	ppm/°C
Line Regulation	$\Delta V_{OUT}/\Delta V_{IN}$	1.8V ≤ V_{IN} ≤ 5.5V		50	400	µV/V
Load Regulation	$\Delta V_{OUT}/\Delta I_{OUT}$	0 ≤ I_{OUT} ≤ 1mA		90	700	µV/mA
		-100µA ≤ I_{OUT} ≤ 0		2	9	µV/µA
Short-Circuit Current	I_{SC}	Sourcing to GND		3		mA
		Sinking from V_{IN}		6		
Long-Term Stability	ΔV_{OUT}/Time	1000hrs at T_A = +25°C		100		ppm

Table 4.13. *Aging parameters of a voltage reference*

After 1,000 hrs, the maximal voltage drift is 100 ppm/year at 25°C. The temperature drift, which is 50 ppm/°C, should also be taken into account. Strictly speaking, line control should also be considered (the variation of the reference voltage when the input voltage varies within the acceptable range) and also the charge control. The advantage of these components is that generally the drift in time is concave. A run-in technique can then be applied by the equipment manufacturer in order to have a maximum drift of the reference voltage. This method is presented in detail in Chapter 8.

4.4. Summary of components with limited service life

Types of components	RDF 2000	FIDES	Manufacturers	Associated parameters	Activation energy
Opto-couplers	Yes	Yes	No	Current, temperature	0.4 eV
LED	Yes	Yes	Yes	Current, temperature	0.4 eV
Wet electrolytic capacitors	Yes	Yes	Yes	Temperature Number of charges/discharges	0.4 eV
Electromechanical relays	Yes	Yes	Yes	Number of on/offs	NA
Interrupters/switches		Yes	Yes	Number of switches	NA
Batteries	NA	Yes	Yes	Temperature, number of charging/discharging cycles	0.53 eV

Types of components	RDF 2000	FIDES	Manufacturers	Associated parameters	Activation energy
Fans	NA	Yes		Temperature	0.21 →0.46 eV
Connectors	Yes	No	Yes	Number of insertions/disinsertions	NA
Power transistors in splitting	Yes	No	No information	Number of power cycles	NA
NAND Flash memory	No	No	Yes	Temperature Number of read/write cycles Data retention	1 eV
Potentiometers	No	No	Yes	Number of rotations	NA
Quartz oscillators	No	No	Yes	Time and temperature	
Voltage regulator	No	No	Yes	Time and temperature	

Table 4.14. *Summary of components assumed to have limited service life*

5

Analysis of Product Performances

Product performance can be analyzed using several types of analysis depending on the phase of the lifecycle.

5.1. Analyses during the design stage

5.1.1. *Worst-case analysis*

Given a function f of "n" real variables xi, i ∈ [1; n]. The differential of this function denoted by "df" is given by:

$$df = \sum_{i=1}^{n} \frac{\delta f}{\delta x_i} . dx_i \qquad [5.1]$$

The worst-case tolerance of this function is defined by:

$$\Delta f = \left|\frac{df}{f}\right| \qquad [5.2]$$

Equations [5.1] and [5.2] lead to:

$$\Delta f = \frac{\sum_{i=1}^{n} \left|\frac{\delta f}{\delta x_i}\right| . dx_i}{f}$$

or

$$\Delta f = \sum_{i=1}^{n} \left|\frac{\delta f}{\delta x_i}\right| . \frac{dx_i}{f} . \frac{x_i}{x_i}$$

or, finally

$$\Delta f = \sum_{i=1}^{n} S_f^{x_i} . \Delta x_i \qquad [5.3]$$

where $S_f^{x_i}$ represents the sensitivity of function f of variable xi and is defined by:

$$S_f^{x_i} = \frac{x_i}{f} . \left| \frac{\delta f}{\delta x_i} \right|$$

Consider the case of a voltage generated by a direct reference voltage Ve and a resistive bridge R1 and R2 according to Figure 5.1.

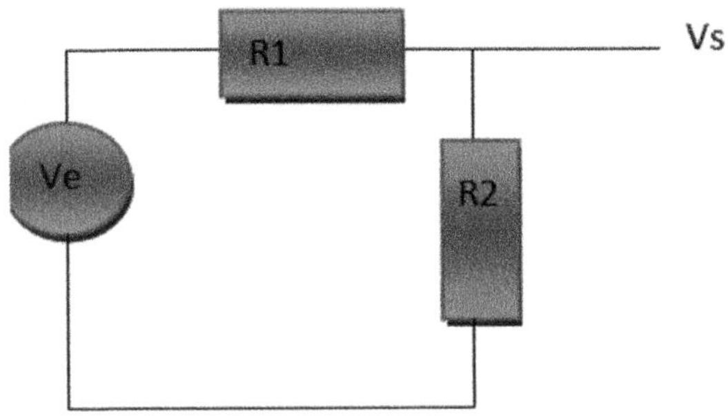

Figure 5.1. *Diagram of a resistive bridge*

Then:

$$Vs = \frac{R2.Ve}{R1+R2} \qquad [5.4]$$

What is the value of ΔVs depending on the tolerances on R1, R2 and Ve?

Based on equations [5.3] and [5.4], the following can be written:

$$\Delta f = S_{Vs}^{R1} . \Delta R1 + S_{Vs}^{R2} . \Delta R2 + S_{Vs}^{Ve} . \Delta Ve$$

Moreover:

$$S_{Vs}^{R1} = \left| \frac{R1}{Vs} . \frac{\delta Vs}{\delta R1} \right| = \frac{R1}{Vs} . \left| - \frac{R2.Ve}{(R1+R2)^2} \right| = \frac{R1}{\frac{R2.Ve}{R1+R2}} . \left(\frac{R2.Ve}{(R1+R2)^2} \right)$$

or

$$S_{Vs}^{R1} = \frac{R1}{R1 + R2}$$

$$S_{Vs}^{R2} = \left|\frac{R2}{Vs} \cdot \frac{\delta Vs}{\delta R2}\right| = \frac{R2}{Vs} \cdot \left(\frac{(R1+R2).Ve - R2.Ve}{(R1+R2)^2}\right) = \frac{R2}{\frac{R2.Ve}{R1+R2}} \cdot \left(\frac{R1.Ve}{(R1+R2)^2}\right)$$

and furthermore

$$S_{Vs}^{R2} = \frac{R1}{R1 + R2}$$

$$S_{Vs}^{Ve} = \left|\frac{Ve}{Vs} \cdot \frac{\delta Vs}{\delta Ve}\right| = \frac{Ve}{Vs} \cdot \frac{R2}{R1 + R2} = \frac{Ve}{\frac{R2.Ve}{R1 + R2}} \cdot \frac{R2}{R1 + R2} = 1$$

Hence:

$$\Delta Vs = \frac{R1}{(R1 + R2)} . \Delta R1 + \frac{R1}{(R1 + R2)} . \Delta R2 + \frac{R1}{R2} . \Delta Ve$$

or

$$\Delta Vs = \frac{R1.(\Delta R1 + \Delta R2)}{R1 + R2} + \Delta Ve \qquad [5.5]$$

NOTE.–

Often, $\Delta R1 = \Delta R2 = \Delta R$, so that equation [5.5] can be written as:

$$\Delta Vs = \frac{2.R1.\Delta R}{R1 + R2} + \Delta Ve \qquad [5.6]$$

Numerical application

Ve = 5V, ΔR = 1% and ΔVs = 2%; the aim is Vs = 3V. A choice is made for R1 = 6,800 and R2 = 10,000.

Hence, Vs = (5.10000)/(10,000+6,800) or Vs ~ 2.976V.

The worst-case value of the tolerance on voltage Vs is then given by:

$$\Delta Vs = \frac{2.6800.1\%}{6{,}800 + 10{,}000} + 2\% \sim 2{,}81\%$$

5.1.2. *Quadratic analysis*

In some cases, worst-case analysis may lead to results that are not industrially feasible. For the sake of clarity, consider the example of an electrolytic capacitor operating as "energy storage". Its size is designed to "cover" the potential brownouts of a power grid. Consider a 6W DC/DC converter supplied by a voltage Ve = 24V, as illustrated in Figure 5.2.

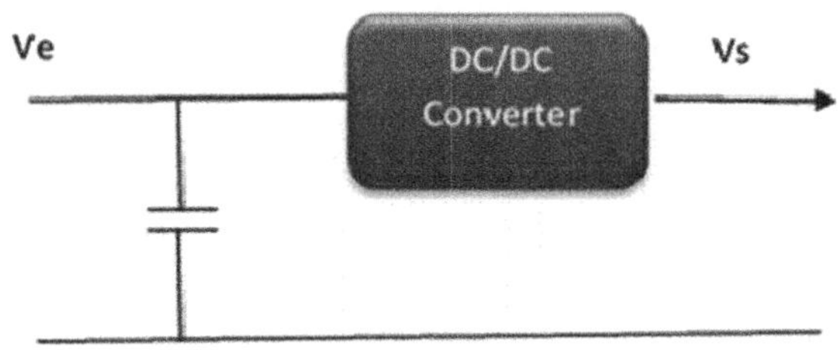

Figure 5.2. *Example of quadratic analysis*

It is important to note that the converter provides the voltage Vs when the input voltage is greater than or equal to a voltage VeMin = 9V. Moreover, the capacitor must be sized to cover the brownouts of Tt = 8 ms. The value of the capacity is then defined by:

$$\frac{1}{2}.\mathrm{C}.(\mathrm{Ve}^2 - \mathrm{VeMin}^2) = \mathrm{P}.\mathrm{Tt} \qquad [5.7]$$

which yields:

$$\mathrm{C} = \frac{2.\mathrm{P}.\mathrm{Tt}}{(\mathrm{Ve}^2 - \mathrm{VeMin}^2)} \qquad [5.8]$$

The numerical value is C ~ 194 µF. Therefore, a capacitor of 220 µF at 35 V could be chosen.

Nevertheless, the transparency of the converter to microcuts requires taking into account the tolerance of the capacitance ΔCc and its evolution (decrease) in time and with temperature ΔCt. Hence, a worst-case analysis would yield:

$$\Delta C = \Delta Cc + \Delta Ct \qquad [5.9]$$

For this type of capacitor, the tolerance guaranteed by the manufacturer is -20% and the time drift is -30%. The worst-case reduction in capacity would therefore be 20% + 30%, hence 50%. Consequently, two 220 µF capacitors would be needed, which could lead to problems in terms of weight and overall external dimensions.

Hence, in these specific cases, quadratic analysis is preferred, and its definition is:

$$\Delta C = \sqrt{\Delta Cc^2 + \Delta Ct^2} \quad [5.10]$$

According to this approach, a single capacitor of 270 μF is required.

5.1.3. *Monte-Carlo analysis*

A more realistic approach involves a less restrictive type of analysis than the worst-case method, and this can be achieved by implementing a Monte-Carlo approach. The variable of interest is then modeled by a random variable according to a given probability distribution. Monte-Carlo simulation involves the random "draw" of values of the variable of interest according to the probability distribution considered. Often, 1,000 random values are a good compromise between accuracy on the bounds of the interval of values obtained, and the simulation calculation time.

5.1.3.1. *Careful approach*

Let us assume that resistance values can be represented by a uniform normal distribution. If the variable xi has a tolerance Δxi, then the uniform distribution is defined by:

$$[x_i - \Delta x_i \, ; \, x_i + \Delta x_i] \quad [5.11]$$

The variable xi, that is, the value of a key characteristic of a component (e.g. the rated value R of a resistance), has an identical occurrence probability throughout the interval $[x_i - \Delta x_i; \, x_i + \Delta x_i]$.

5.1.3.2. *Realistic approach*

In reality there is of course no reason why the resistance values in the respective interval should be equiprobable. Let us then assume that the resistance values can be represented by a normal distribution of parameters μ and σ. Given the interval of possible values at ± 6σ, almost 100% of the interval of possible values of the variable xi is represented. Therefore, a value of the variable xi belongs to the following interval:

$$[\mu - 6.\sigma \, ; \, \mu + 6.\sigma] \quad [5.12]$$

The mean value μ is directly equal to the rated value of the key characteristic (e.g. the rated value R of a resistance). If the accuracy of resistance R provided by the manufacturer is "p%", then the value of the standard deviation σ can be estimated. For a better understanding, Figure 5.3 illustrates the method.

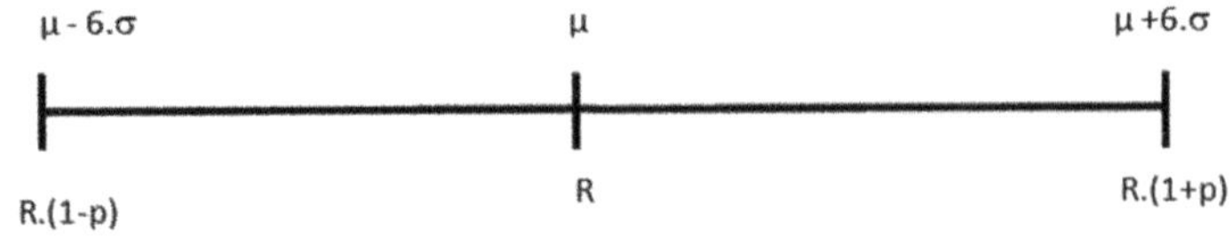

Figure 5.3. *Estimation of the standard deviation of the normal law for Monte-Carlo simulation*

Based on the previous figure, by identifying the extreme values of the considered interval, the following equation is deduced:

$$\sigma = \frac{\mathbf{R.p}}{\mathbf{6}} \qquad [5.13]$$

5.1.3.3. *Example of the resistive bridge*

If a Monte-Carlo simulation is used under the hypothesis of uniform distribution, 2.7% accuracy is obtained for voltage Vs, which is very close in this example to the worst-case value of 2.81%. Using a normal distribution leads to accuracy of approximately 1%.

R1	R2	Vcc	Vs	dVs
6741	9951	4,932	2,94023077	1,21%
6824	9928	5,049	2,99226791	-0,54%
6849	9996	5,041	2,99138237	-0,51%
6815	9923	5,010	2,97014159	0,20%
6800	9945	5,027	2,98557868	-0,32%
6830	10093	5,069	3,02318838	-1,58%
6737	10093	5,053	3,03029881	-1,82%
6832	10009	5,063	3,00905926	-1,10%
6861	10006	5,004	2,96851983	0,26%
6856	9916	4,936	2,91827904	1,95%
6785	9928	5,084	3,0200414	-1,47%
6855	9929	5,014	2,9661586	0,34%
6793	9949	5,079	3,01821592	-1,41%
6740	9967	5,069	3,02404519	-1,61%
6794	10089	4,916	2,93771984	1,29%
6816	10037	5,078	3,02426191	-1,62%
6766	10031	4,980	2,97400607	0,07%
6825	10013	4,952	2,94479012	1,06%
6866	10051	5,070	3,01226991	-1,21%
6807	10012	4,948	2,94544123	1,03%
6833	9985	5,065	3,0071367	-1,04%
6769	10069	5,095	3,04677248	-2,37%

R1	R2	Vcc	Vs	dVs
6 781	9 974	5,00	2,9766	-0,01%
6 820	9 993	5,00	2,9746	0,05%
6 799	9 984	4,99	2,9677	0,28%
6 783	10 017	5,01	2,9887	-0,42%
6 822	10 036	5,03	2,9918	-0,52%
6 798	9 980	4,99	2,9690	0,24%
6 818	10 006	5,02	2,9874	-0,38%
6 810	10 019	5,00	2,9776	-0,05%
6 787	10 019	5,00	2,9799	-0,13%
6 793	10 025	4,99	2,9734	0,09%
6 804	10 000	5,01	2,9786	-0,08%
6 801	10 041	4,99	2,9738	0,08%
6 818	10 011	5,00	2,9746	0,05%
6 792	9 990	5,01	2,9814	-0,17%
6 804	10 019	4,99	2,9748	0,05%
6 797	10 006	5,01	2,9859	-0,33%
6 809	9 985	4,98	2,9638	0,42%
6 802	9 997	5,03	2,9908	-0,49%
6 806	9 995	5,00	2,9725	0,12%
6 799	9 999	4,97	2,9593	0,57%

Table 5.1. *Example of simulation for the uniform law (on the left) and for the normal law (on the right)*

5.1.4. *Numerical simulations*

There are different types of simulations for a product. The advantage of these simulations is that they can be used to develop a more mature product without having to build a prototype. There is a rich literature on electronic, thermal and mechanical simulations, which cannot realistically be covered in a single chapter. Therefore, we only present here the aspects that we consider important within the scope of this book.

Indeed, thanks to electronic simulations of components, it is possible to check certain parts of the electronic design using specific software, such as Pspice (Tuinenga 1997) and the libraries of models of components provided by the manufacturers.

As for thermal simulations, thanks to finite element techniques, it is also possible to estimate the temperature of each component, as illustrated in Figure 5.4.

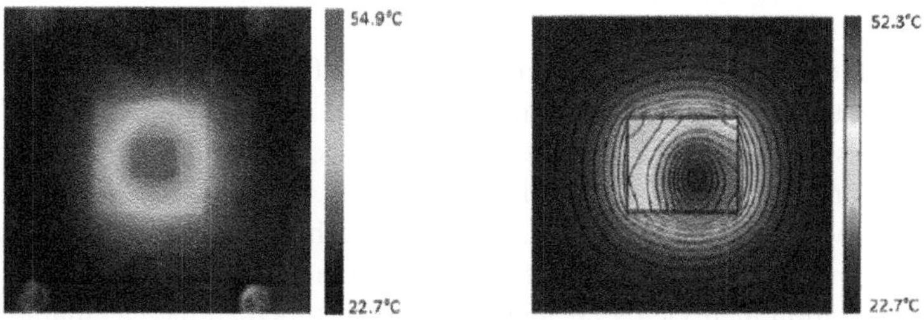

Figure 5.4. *Comparison between thermal simulation and measurement. For a color version of this figure, see www.iste.co.uk/bayle/maturity1.zip*

Temperature estimates are very important, either for derating analysis or predictive reliability estimation. Consider the example of a MOSFET in Dpak package dissipating a power Pd =1.5 W, at an ambient temperature θa = 45°C.

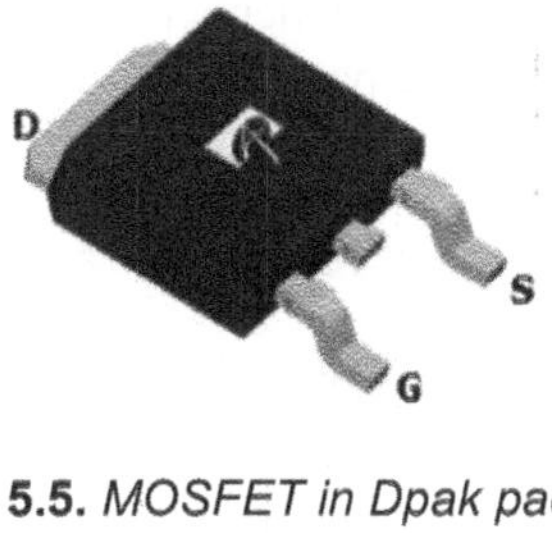

Figure 5.5. *MOSFET in Dpak package*

To estimate the MOSFET junction temperature, the thermal Ohm's law is written as:

$$\theta j = \theta a + \text{Rthja.Pd} \qquad [5.14]$$

The datasheet for components of the MOSFET includes the following data:

Thermal Characteristics					
Parameter		**Symbol**	**Typ**	**Max**	**Units**
Maximum Junction-to-Ambient [A,G]	$t \leq 10s$	$R_{\theta JA}$	15	20	°C/W
Maximum Junction-to-Ambient [A,G]	Steady-State		41	50	°C/W
Maximum Junction-to-Case [D,F]	Steady-State	$R_{\theta JC}$	2	2.4	°C/W

Considering the standard value, the resulting junction temperature is: θj = 45 + 41*1.5 = 106.5°C. However, the index G informs us that these tests are performed with the device mounted on 1 in^2 FR-4 board with 2oz. Copper, in a still air environment with T_A=25°C.

To put it differently, the thermal resistance of 41°C/W is for a printed FR4 circuit with a transfer surface of 1 square inch, and a copper thickness of 2 oz. This never happens in practice. There are often signal or ground layers underneath the side of the printed circuit where the component is mounted. Furthermore, it is possible to have heat sinks, forced ventilation, or a heater for other types of package (e.g. TO220 and TO247) so that the thermal resistance value is significantly smaller. Thermal simulation then yields far more realistic results than the calculation based on data from the datasheet. Continuing with the example of MOSFET with a thermal resistance of 15°C/W, the resulting junction temperature is θj = 45 + 15*1.5 = 67.5 °C, which is nearly 40°C lower.

For the mechanical simulation, the situation is more complex, primarily because component manufacturers provide no vibration information. Indeed, this position is not really objectionable since the mechanical strength of the component depends largely on its mechanical design in the product. For example, the loops of the electronic card in which the component is inserted strongly depend on the characteristics of the card and of the package into which it is integrated.

5.2. Analyses during the manufacturing stage

This type of analysis relies on statistical process control (SPC), a tool widely used in the manufacturing field for process control. Control during manufacturing contributes to stable production with a minimum number of products that do not comply with specifications. SPC aims to achieve higher quality using statistical tools (reception control, experimental drawings, regression techniques, Pareto diagrams, control charts, etc.), which can lead to more centralized, less dispersed production.

There is a rich and abundant literature on this subject, and hence, it cannot be covered in a single chapter. Any interested readers are invited to refer to the literature for further details.

6

Aggravated Tests

There are many articles and books that offer very precise details of aggravated tests; this is beyond the scope of this chapter. Although a concise description of these tests is provided here, the fundamental aim of this chapter is to explore how these tests relate to product maturity.

6.1. Definition

In the context of this chapter, aggravated tests can be defined as follows:

> Tests that submit the product or some of its constituent elements to (climatic, electrical, mechanical) constraints that are progressively increased to values significantly higher than those specified, within the product operation and/or destruction limits.

According to the manufacturers, the above strict definition of aggravated tests does not cover all these practices, particularly in terms of scale of application (the product), application technique (no bias on the value of constraints seen by the product) and even final purpose of the tests.

6.2. Objectives of aggravated tests

There are numerous articles, documents, presentations, etc. on aggravated tests, which include the following points on the use of conducting aggravated tests (McLean 2009):

– improve product robustness by applying quite serious stresses;

– highlight the potential weaknesses inherent to product design and to the technologies involved;

– estimate design margins, as illustrated below;

– at the card or subassembly level, make a selection based on various technologies or various providers and choose the most robust.

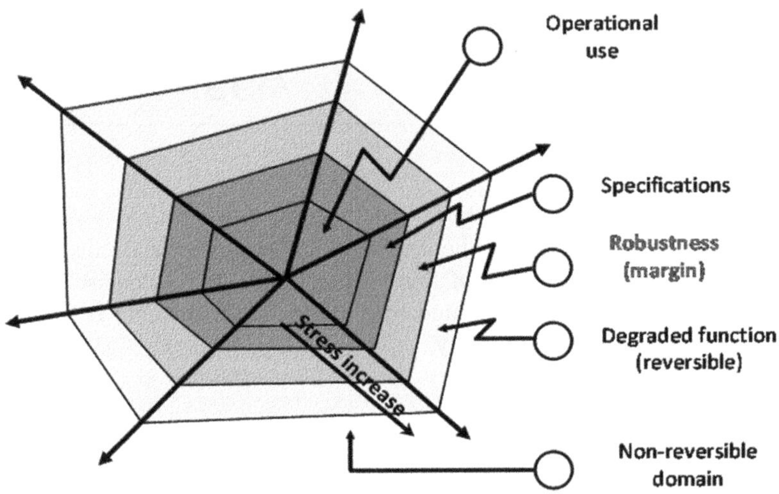

Figure 6.1. *Illustration of the principle of aggravated tests. For a color version of this figure, see www.iste.co.uk/bayle/maturity1.zip*

The above points are evidently correct and not to be questioned. However, the main objective of aggravated tests is to participate in building product maturity. Estimating the operational limits of the product (and deducing robustness margins), eliminating design weaknesses, etc. are actually means of contributing to building product maturity.

Indeed, identifying design failures and then correcting them will notably improve product reliability by suppressing the reliability increase stage observed in the first operational years of the product, reducing potential aging failures during or by the end of product operation. In other terms, aggravated tests contribute significantly to rendering the operational reliability of the product as constant as possible. Moreover, the resulting robustness margins will also reduce the influence of accidental overloads (catastrophic failures) that are generally present during product operation. This will generally have a non-negligible influence on the observed reliability level, as illustrated in Figure 6.2 below.

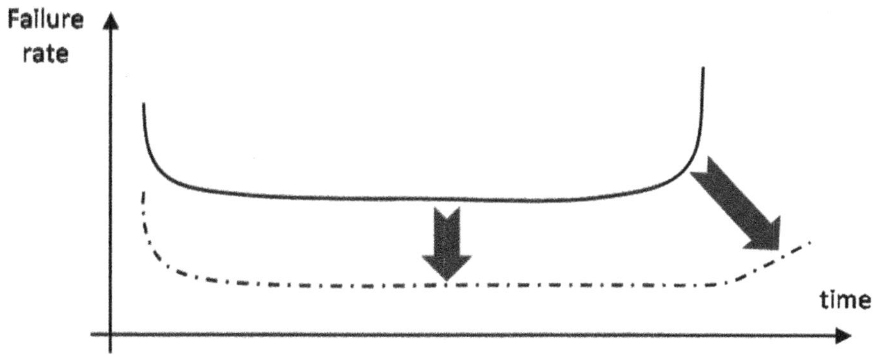

Figure 6.2. *Influence of aggravated tests on product reliability*

6.3. Principles of aggravated tests

The principles of aggravated tests are:

– to subject the product to various environmental stresses by progressively increasing the levels step by step until the occurrence of a failure;

– to conduct a root cause analyses for each failure in order to identify whether it is a design defect/failure due to:

- a design problem,

- a technological limitation;

– to decide whether or not corrective action is needed;

– to reiterate the process of elimination of these weaknesses until design margins that are considered sufficient are obtained.

The "guiding principle" of an aggravated test is not a willingness to "comply" with the requirements of a specification. Its objective is not the verification of technical performance, nor an estimation of service life. The "guiding principle" of an aggravated test is to bring a product to the limit of its capacities, or even to break down, by subjecting it to "what is needed, where it is needed, when it is needed".

In order to choose the strategy that best fits the manufacturing context of interest, two essential questions must be addressed, namely:

– Is an aggravated test necessary?

– If so, what is the most relevant type of test?

In order to answer these questions, all the constraints imposed on the equipment manufacturer must be analyzed.

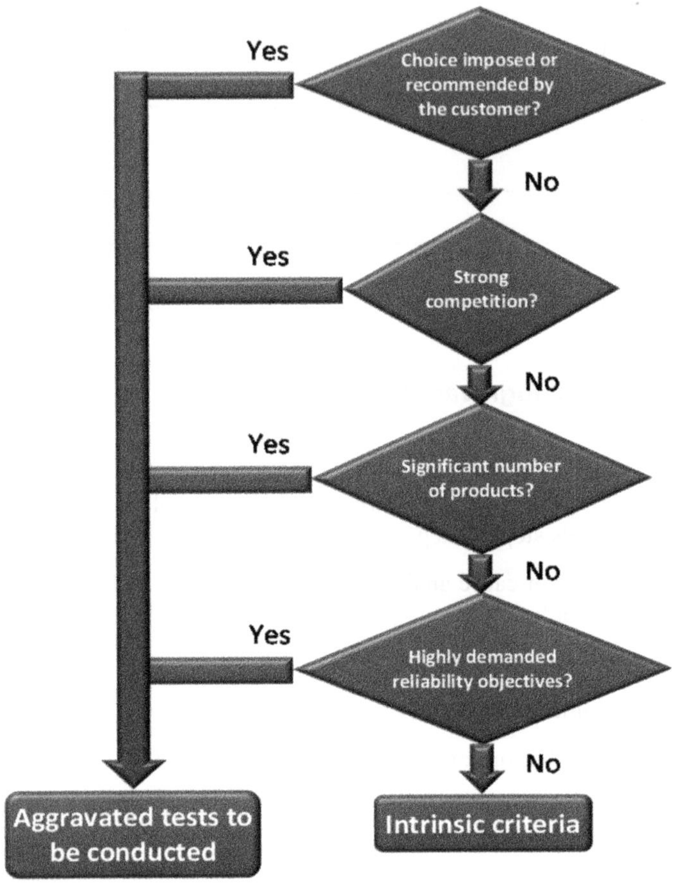

Figure 6.3. *Examples of extrinsic decision criteria for conducting aggravated tests. For a color version of this figure, see www.iste.co.uk/bayle/maturity1.zip*

There are two types of aspects guiding the choice of the method:

1. Constraints **extrinsic** to the equipment manufacturer:

- recommended or even imposed by the system manufacturer;

- a strict life profile in terms of level of physical contributions. In this case, we can consider that the verification of the proper design and the estimation of design margins are major elements;

- the context of a highly competitive market, in which the other manufacturers systematically conduct this type of test;

- very important dependability specifications (critical dreaded events, carrier availability);

- a large number of products to be delivered. Indeed, as already noted, this type of test aims to eliminate potential design weaknesses. It is known that this type of failure is very often reflected in premature aging that is observed throughout the batch of delivered products. The design modification will therefore have a significant financial impact;

- very important brand image, guaranteeing the sales of other types of products. Failures observed during operation involve a risk of significant damage to the brand image of the equipment manufacturer.

Figure 6.3 gives the paths of extrinsic decision criteria, which can be adapted as needed. In Figure 6.3, the given criteria can be interpreted as follows:

– aggravated tests are recommended by the system manufacturer. Indeed, when the latter has a reference product for which this type of test is recommended, it would be difficult not to follow this recommendation;

– if competition is strong, then reliability (or quality) of the equipment should be maximal and aggravated tests could contribute to fulfilling this need;

– if the number of products is significant, then the investment required for the aggravated tests will be returned more quickly (and if the product is small, several can be placed in the equipment being used, reducing the time required to conduct the tests);

– if reliability objectives are high, then aggravated tests are a way of addressing this need by systematically detecting failures and methodically repairing them.

2. Constraints **intrinsic** to the equipment manufacturer:

- a strategy of conducting tests that are recommended or even imposed by the company;

- a large variety of such products that are specific to the equipment manufacturer;

- the company's brand image;

- an equipment manufacturer having a "product policy", meaning operational blocks that are introduced in all the types of products designed by the equipment manufacturer. In case of a design failure on an operational block, all the products designed will feature this failure, with disastrous results;

- the impact of such tests on the cost of products. Indeed, in certain domains, the cost of such tests may have a significant impact on the final cost of the product and may potentially lead to the loss of important markets;

- the type of manufacturing domain (aeronautics, automotive, domestic appliances, etc.). This question is not applicable to certain industrial domains, in which aggravated tests must be conducted (e.g. the space industry, which does not tolerate the slightest error).

Figure 6.4 gives the paths of intrinsic decision criteria, which can be adapted as needed.

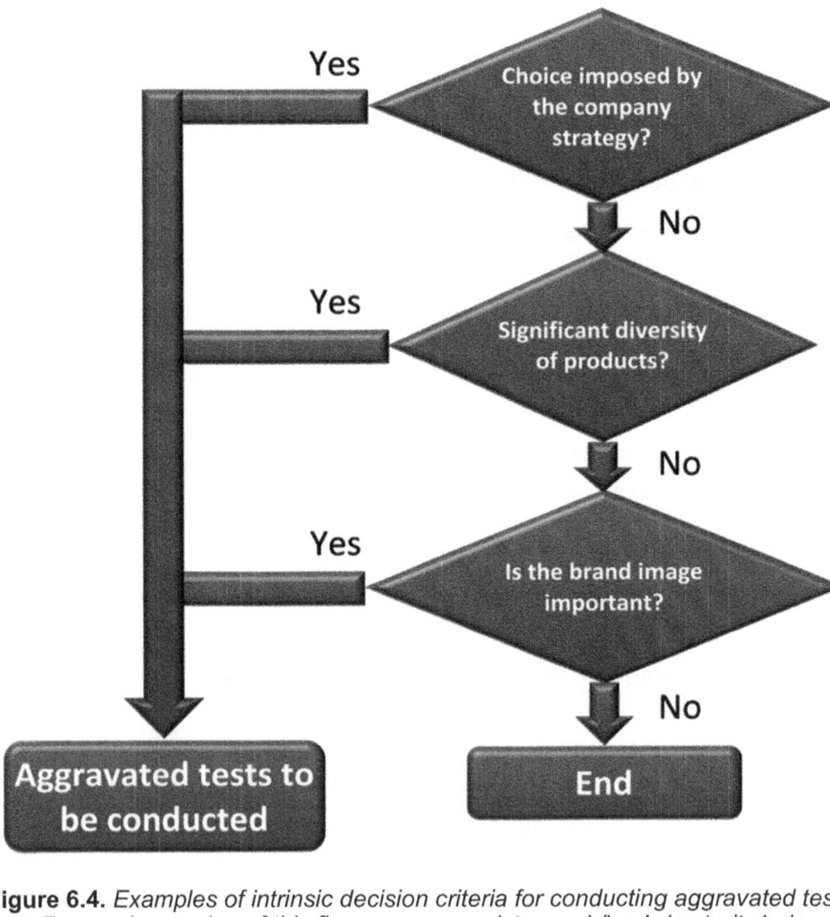

Figure 6.4. *Examples of intrinsic decision criteria for conducting aggravated tests. For a color version of this figure, see www.iste.co.uk/bayle/maturity1.zip*

Although in certain cases it may be possible to use other aggravated test methods, the recommended method is HALT/HASS (McLean 2009), which is the focus of the next section.

6.3.1. *Choice of physical constraints*

All previously presented advantages of aggravated tests are not real unless the product is correctly stressed. The choice of physical contributions is therefore essential. Several studies were conducted (IES 1984) on this subject, and their results can be summarized in Figure 6.5 (below).

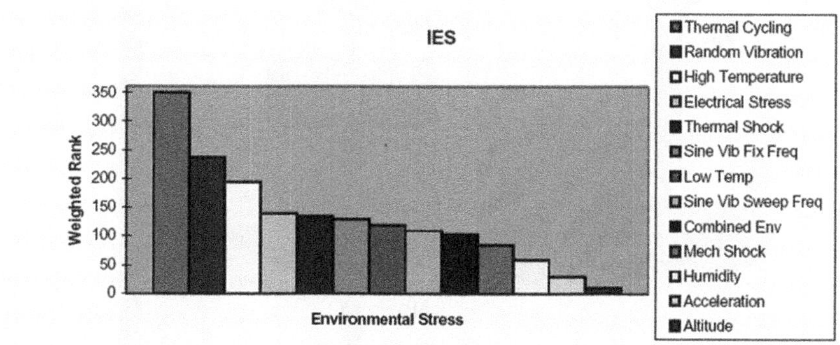

Figure 6.5. *Comparison of the influence of physical contributions for aggravated tests. For a color version of this figure, see www.iste.co.uk/bayle/maturity1.zip*

As it is impossible to simultaneously generate constant temperature and have thermal cycling, the dominant condition is kept, namely, the thermal cycling. In fact, the most important physical contributions (thermal cycling, vibration and power cut) were grouped in a single machine, greatly simplifying these tests, particularly in terms of time and costs.

6.3.2. *Principle of HALT*

In order to find the operational limits of a product, the approach illustrated in Figure 6.6 is used.

Here, the technological limit is identified as being the maximal values guaranteed by the manufacturer of the components. For example, for a component that is guaranteed up to +85°C, the technological limit of the product will be reached when the temperature level of the aggravated test exceeds this value, even if the product is still fully operational.

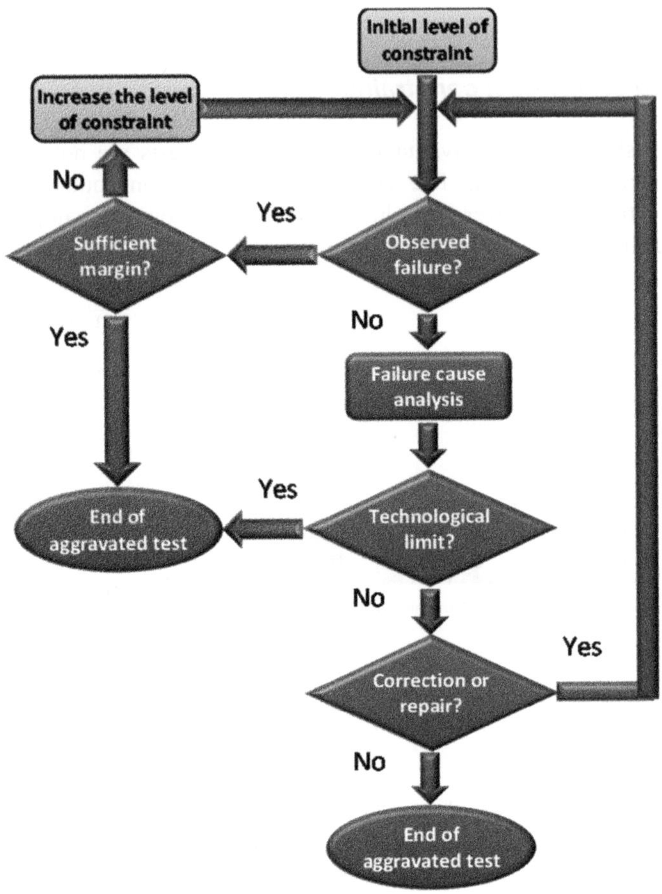

Figure 6.6. *Principle of HALT tests. For a color version of this figure, see www.iste.co.uk/bayle/maturity1.zip*

HALT testing involves the following process:

– finding the low-temperature operational limit (LTOL). The starting point is the ambient temperature, which is then decreased in steps of generally 10°C for 10 min. If the product is operational, the temperature is further decreased by 10°C until an operational failure occurs. This can be illustrated by Figure 6.7 (below);

– finding the high-temperature operational limit (UTOL). The starting point is the ambient temperature, which is then increased in steps of generally 10°C for 10 min. If the product is operational, the temperature is further increased by 10°C until an operational failure occurs. This can be illustrated by Figure 6.8 (below).

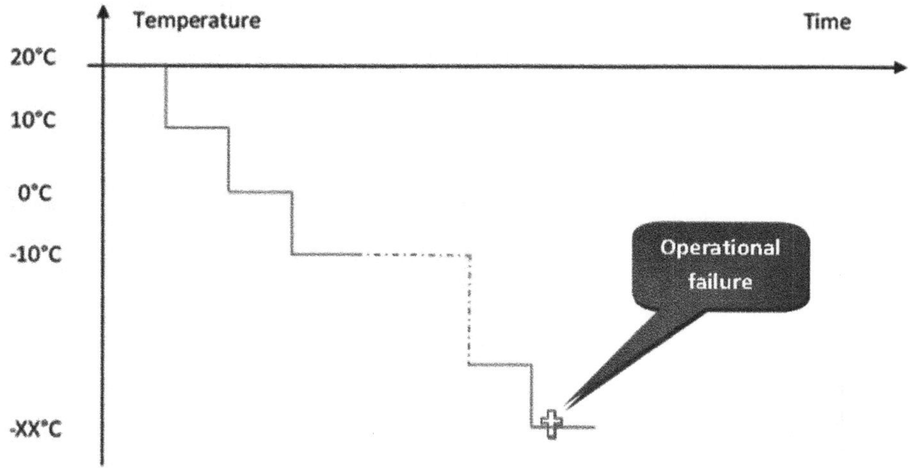

Figure 6.7. *Finding the low temperature operational limit*

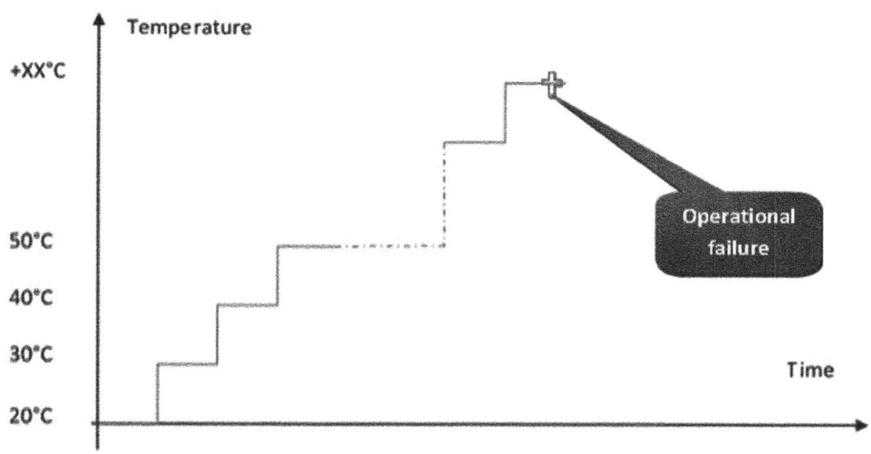

Figure 6.8. *Finding the high temperature operational limit*

NOTE.– It is not surprising that destructive low-temperature limits are rarely reached by electronic materials. They are indeed more sensitive to high operational temperatures than to low operational temperatures.

– Finding the vibration operational limit (UVOL). The starting point is the vibration in Grms, which is increased in steps of generally 5 Grms for 10 min. If the product is operational, the vibration level is further increased by 5 Grms until an operational failure occurs. This can be illustrated by Figure 6.9 (below).

It is important to note that vibrations take place along 6 degrees of freedom (one translation and one rotation per axis) and that the product being tested is at the ambient temperature. The duration of one step is of minimum 10 minutes, but it may take longer to conduct the required tests. Vibrations generally range between 2 Hz and 5 kHz.

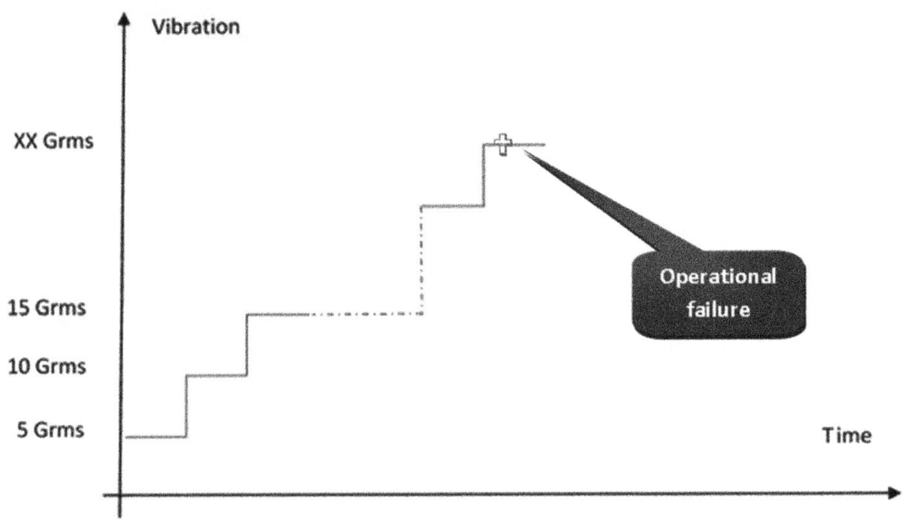

Figure 6.9. *Finding the vibration operational limit*

– Finding the operational limits by a combination of vibrations and thermal cycling.

A combination of constraints, such as vibrations and thermal stresses, can be used to reveal latent weaknesses of the equipment, which do not occur under a single constraint (e.g. temperature only or vibration only).

NOTE.– Moreover, during each of these steps, the package is monitored (controlling the operational tolerance drift and recording the constraints and their responses at the card level) so that defects due to temperature change are detected.

– Unless it is checked by other tests, it is useful to check the cold start of the equipment at various steps. This is an additional test, which involves checking the start of equipment under stabilized low temperatures, in order to determine its limit. It is distinct from ON/OFF cycles, as the latter are conducted on an initially operating equipment, whose components have a higher temperature than the environment, especially if it is very dissipative.

Figure 6.10 illustrates this process.

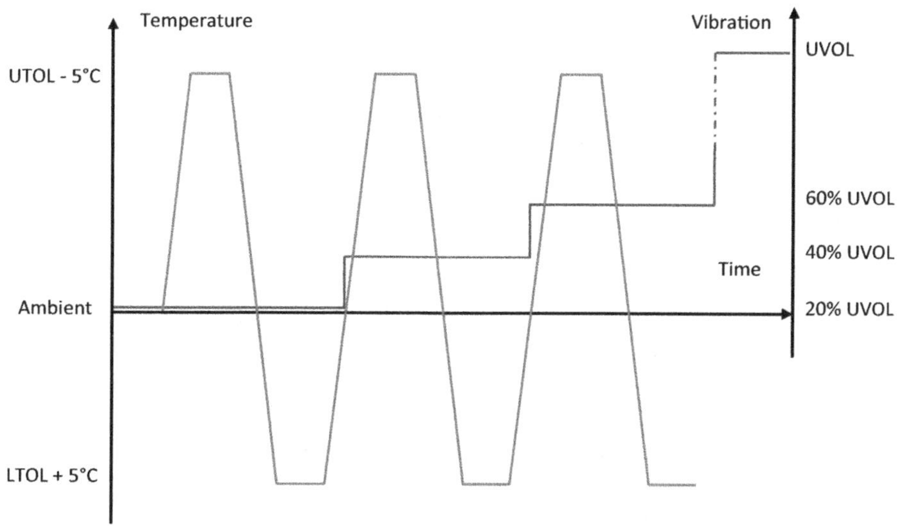

Figure 6.10. *Finding the operational limit for combined thermal cycling/vibration constraints. For a color version of this figure, see www.iste.co.uk/bayle/maturity1.zip*

NOTE.– The above example of combined cycle does not take into account thermal inertias, which can be generated by a collapse. The thermal profiles (temperature slope/step duration) must be corrected to fit the constraints. The time needed to initialize the equipment and the monitoring software must also be taken into account.

– The application tolerances of the acceptable constraints are generally of ±3dB in vibration about the rated profile and ±2°C in thermal cycling.

6.3.3. *Specific or additional constraints*

The nature of the physical contributions or specific stresses to be applied is suited to the tested equipment. It is therefore difficult to formulate a generic list. Moreover, when non-electronic equipment is involved, the constraints and the stresses to be applied through a testing process are defined case by case.

For electronic materials, the following specific tests can be taken into consideration during a testing process, if their relevance is approved during the stage of constraint selection:

– power cycle;

– cycle of various supply voltages combined with previous sequences;

– finding the limit alternative voltages as well as the limit frequencies;

– finding the maximum direct supply voltages.

For mechanical materials, the following constraints can be taken into consideration:

– continuous increase of mechanical load;

– increase of speed or frequency of mechanical loads;

– incremental increase of the load acceleration.

The above list is just an example, and is not exhaustive. Its purpose is to show the diversity of constraints that can be implemented during an aggravated testing process. The preliminary stage of analysis and determination of constraints and constraints to be implemented during a process is therefore very important.

6.3.4. *Number of required samples*

According to the referenced literature (McLean 2009), due to the dispersion of results according to the tested products and the representativeness, it is recommended to use four products per robustness test. However, for highly expensive products and provided that the dispersion of the various operational limits observed is very weak, a single sample may be used per type of test.

6.3.5. *Operational test, diagnosis and identification of weaknesses*

It should be possible to stimulate and test the product throughout the testing process so that reversible malfunctions and breakdowns are detected. Test coverage should be sufficient, and it should be possible to observe the evolution of temperature performances, if the function depends on it. It is highly recommended that we conduct a preliminary analysis of the range of temperatures specified for all the components and their expected performances according to the datasheet provided by the manufacturer. This will help in identifying the operational chains that will probably be impacted first during temperature margin tests and to make sure that the test covers these functions.

6.3.6. *Monitoring specification*

The set of parameters to be measured and therefore to be monitored during the various testing sequences should be defined before the HALT tests. Indeed, it may prove necessary to adapt certain testing methods and even the equipment itself in order to have or allow a proper monitoring.

It is important to differentiate between:

– Operational monitoring, which is the monitoring of product functionalities. Operational monitoring of the equipment is required for deciding on the proper operation of the electronics at any moment of the aggravated test. Defects can be evidenced by monitoring analysis, and it is important to know which operational constraint generated which defect. This makes it possible to identify the electronic functions that failed under constraint and to rapidly define the failing component.

– Recording the stresses perceived by the equipment: a proper instrumentation is a must for the analysis of resulting defects, notably the identification of defects due to reaching the technological limits.

For each test monitoring sequence, the following must be defined:

– parameters to be monitored;

– operational constraints to be applied;

– functions covered by the tests;

– frequency of tests (every 30 seconds, etc.);

– type of test (intrusive, such as the injection of breakdowns, specific or mobile probe);

– number of defects based on which a breakdown is confirmed;

– inhibition conditions;

– behavior of the associated product.

6.3.7. *Instrumentation*

In order to properly grasp the levels of involved constraints, the product subjected to aggravated tests must be instrumented. This instrumentation (thermocouples, accelerometers, constraint meters, etc.) can also be used to instantaneously measure the response of the product to constraints. The implementation of this instrumentation is defined based on risk analyses and focuses on the most sensitive components, therefore, the electronic components. For example, on electronic cards, accelerometers are placed on vibration-sensitive components in order to check their maximal displacements during testing.

Vibrational and thermal profiles must be measured at points where it is possible to reconcile the operational defect of the equipment and the applied constraints, or even to estimate the card damage. These profiles must be recorded by accelerometers and temperature probes whose location guarantees a proper application of the constraint during burn-in, in view of proper traceability.

6.3.8. *Root cause analysis, corrective actions and breakdown management*

Root cause analysis is an integral part of the aggravated testing process. It offers the certainty that the malfunction is related to the test during which the defect appeared, and not to previous stresses, or that the defect is not related to anything else (damage related to the rig, manufacturing defect, manipulation error, etc.).

The root cause analysis process must be documented with the failure mode, the exact cause of failure or its suspected cause, in case of uncertainty.

Failure analysis involves two significant stages:

1) electrical diagnosis, which confirms and characterizes the electric defect and guides the physical analysis;

2) physical analysis, which identifies the defect location and determines its root cause.

By identifying the root cause of the failure, it is possible to devise appropriate corrective actions, depending on the margin being found. Depending on the type of weakness detected during the tests and characterized by the analysis, corrective actions may involve:

– changing the reference of a component;

– modifying the location of a component;

– locally reinforcing a component (collage, underfilling);

– reviewing the mechanical design of the card (thermal drain, mechanical maintenance).

A test is considered successful if it highlights as many potential defects of the material as possible. It is important to consolidate all this information, upstream and downstream of the operation, in order to optimize the initially chosen aggravated test. This requires implementing a product to collect information related to failures. This product must be related to the failure modes observed:

– during the aggravated test;

– after the aggravated test (material acceptance, client refusal).

Therefore, the information collection may be composed of:

– the testing reports for each material;

– the results of the aggravated tests of the materials;

– the return of experience (RETEX).

A copy of the testing report is attached to each unit subjected to an aggravated test. It contains at least the following elements:

– identification of the material (designation, reference, serial number);

– the test result.

For each breakdown observed, the following elements should also be mentioned in the report:

– the type of cycle during which the defect occurred (e.g. vibration or thermal cycle);

– the findings;

– the defective components;

– corrective actions;

– the number and type of cycles throughout the stages;

– the date.

It is important to have a follow-up of corrective actions. Hence, a file containing the results of aggravated tests should be edited regularly. It indicates the number of objects subjected to aggravated tests for each type of material and mentions the causes of breakdowns as well as the actions taken in order to repair the material. In principle, this report should be drawn up by the quality assurance in manufacturing department.

Finally, a report of the equipment returns must also be edited regularly. For each piece of equipment returned, the report should indicate:

– the conditions under which the failure occurred;

– the operating mode of the equipment;

– the age of the equipment;

– the operating hours, the number of kilometers, etc.;

– the serial number of the equipment;

– the place and date of the incident;

– the conditions of the return to the workshop (transportation mode, operations conducted, etc.);

– the examination result;

– the cause of failure (if confirmed);

– the failed component(s), if applicable.

These elements can be used in the intervention on design, manufacturing operations or on the effectiveness of aggravated tests.

6.3.8.1. *Failure analyses*

The objective of failure analysis is to determine the failure mechanism and the root cause of the defect. These causes can be grouped into the following categories:

– manufacturing defect of the component, printed circuit, etc.;

– damage induced by the manufacturing cycle, handling errors;

– damage induced by the rig or by the testing means (electrical constraint);

– design defect.

The interest of the analysis is thus to act at the earliest possible instant on the origin of the failure and to conduct the corrective actions (manufacturing processes, burn-in operations) to prevent failure recurrence. It also reinforces the effectiveness of the aggravated tests.

Failure analysis involves two large stages:

– Electrical diagnosis: this is used to confirm and characterize the electrical defect and thus guide the physical analysis.

– Physical analysis: this is most often destructive and can be used to locate the defect and determine its root cause.

Failure analyses can also be used to evaluate the effectiveness of aggravated tests by determining the way in which the applied constraints led to revealing the defect. Failure analysis can be used to distinguish the real from "false" defects. "False" defects are those that can be attributed to a component during the card or equipment diagnosis, though they are not directly due to the component (or another element).

As an illustration, the typical average cost of such an analysis ranges between approximately 4,000 € and 10,000 € in a specialized laboratory. The failure analysis cost depends on:

– the complexity of the component (test and electrical diagnosis);

– the reproducibility of the defect (intermittent breakdown);

– the type of physical defect (ease in locating and identifying the defect).

An example of observed failure/breakdown and corrective action: during the vibration aggravated tests, we observe a failure of component pins. The adequacy of the component robustness is then to be determined according to the operational demands (depending on the margins to be reached).

6.4. Robustness

6.4.1. *Estimation of robustness margins*

This chapter explains how to determine the operational margins of a product by aggravated tests. In fact, many documents recommend the estimation of the robustness margins of a product but they do not explain how to check that they are sufficient. This chapter proposes a method to meet this objective.

Margins are determined once the operational limit is estimated through HALT tests. The margin is then defined as follows:

$$Margin = Operational\ limit - specification\ value$$

For example, at a high temperature, a product can be specified at 85°C and have an operational limit at 108°C. The resulting margin is therefore 108 – 85 = 23°C. This notion of the margin is obviously applicable to all the physical contributions implemented in the aggravated test.

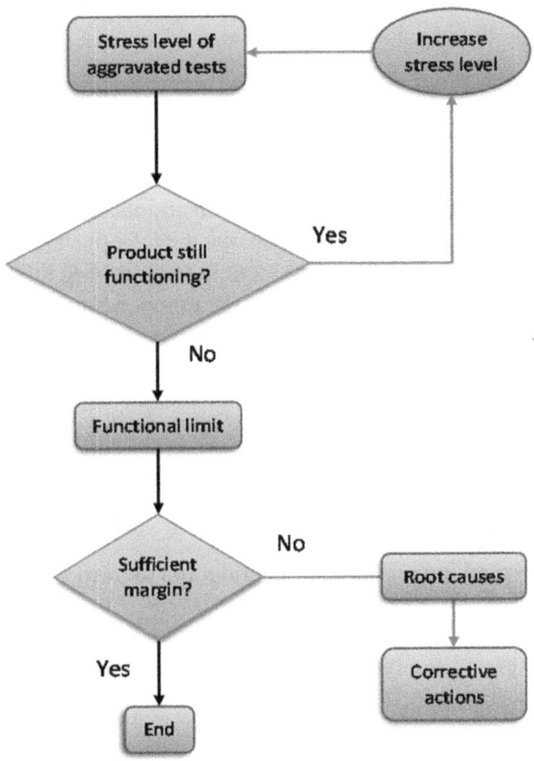

Figure 6.11. *Links between margins and aggravated tests. For a color version of this figure, see www.iste.co.uk/bayle/maturity1.zip*

6.4.2. *Sufficient margins*

The fact that a design should be reviewed when the margins are insufficient is often mentioned. This requires a definition of what a "sufficient margin" means. Moreover, this may depend on the manufacturing constraints inherent to the product concerned, and what is acceptable for one application may not necessarily be so for another.

Hence the use of the "resistance/constraint" method, well known in mechanics, is proposed here. Here, the constraint is the specification required by the system manufacturer, while the resistance is the operational limit estimated during the HALT test.

Resistance R and constraint C are considered as two random variables following a normal distribution, which leads to:

$$\textit{Operational limit} = \mathrm{N}(\mu_r; \sigma_r) \text{ and } \textit{Specification} = \mathrm{N}(\mu_c; \sigma_c)$$

where

μ_c is the level of the physical contribution of interest given in the specification;

σ_c is obviously zero;

μ_r is the average of operational limits of products having undergone a HALT test;

σ_c is the standard deviation of the operational limits of products having undergone a HALT test.

It is possible to estimate failure probability, meaning the probability that the resistance R is weaker than the constraint C, leading to a product failure. The margin is then considered as sufficient when the probability $P(R < C) < P_{target}$, where P_{target} is the objective probability depending on the application. The difficulty is then in estimating the target probability.

Given the previous hypotheses, it can be shown that the probability of having a failure (in fact, that the resistance is weaker than the constraint) is given by:

$$P(R - C < 0) = \phi\left(\frac{\mu_c - \mu_r}{\sqrt{\sigma_r^2 + \sigma_c^2}}\right) \quad [6.1]$$

where ϕ is the distribution function of the standard normal distribution. Since σc is zero, this finally leads to:

$$P(R - C < 0) = \phi(\frac{\mu_c - \mu_r}{\sigma_r}) \quad [6.2]$$

The target probability for the two types of manufacturing applications should therefore be estimated.

6.4.2.1. *The reliability objective is the probability to fulfill the mission*

Considering all the elements involved in the calculation of this probability as independent, if "n" is the number of elements, then:

$$R_p(t) = \prod_{i=1}^{n} R_i(t) \qquad [6.3]$$

Referring to Chapter 1, this probability can also be written in terms of the type of failures as follows:

$$R_p(t) = R_j(t).R_c(t).R_v(t) \qquad [6.4]$$

where R_j, R_c and R_v are the proper operation probabilities of youth, catastrophic and aging failures, respectively. It is possible to formulate a hypothesis according to which, thanks to burn-in, there will be no youth failures or, more reasonably, their influence will be negligible or $R_j(t) \sim 1$. Based on predictive reliability methods, an estimation of the proper operation probability of catastrophic failures can also be made.

Finally, the analysis of components with limited service life, which was developed in this chapter, can be used to give an estimation of the proper operation probability of the aging failures. Hence, based on all these estimations, it is possible to deduce a value of the probability P_{target}. To illustrate the proposed approach, consider the following example:

Assume that a product is specified at a maximal temperature, of +85°C. It can be readily deduced that:

- μ_c = 85°C;
- σ_c = 0.

Assume that the HALT test provided a temperature operational limit of +90°C, or μ_r = 90°C (the failure during the HALT test was observed at Ta = 95°C). If several products were subjected to the HALT test, the standard deviation of the temperature operational limit can be estimated. If a single product was subjected to the HALT test, then a standard deviation of 4°C can be considered, according to (McLean 2009). Let us formulate this last hypothesis so that σ_r = 4°C.

The failure probability given by equation [6.2] can then be calculated, Pdef ~ $8.8\ 10^{-5}$.

Let us assume that the objective of product reliability at the end of mission of duration Tm = 10 years is Rp(Tm) = 95% and that the catastrophic failure rate is $\lambda c = 2.21\ 10^{-8}$.

The proper operation probability of catastrophic failures is therefore at the end of mission Rc(Tm) = 98.083%. Consequently, if there are no youth failures, no aging failures, the probability P_{target} must be:

$$R_{target} \geq \frac{R_{obj}}{R_c(\mathrm{Tm})} \quad [6.5]$$

Or numerically: $R_{target} \geq 96.857\%$. The failure probability was estimated at $8.8\ 10^{-5}$ or a proper operation probability of 99.379%.

The reliability objective is therefore met in this purely virtual example.

6.4.2.2. *The reliability objective is an MTBF*

For this type of application, reasoning directly in terms of proper operation is no longer applicable, since the specified reliability objective is an MTBF. For this type of application, the maintenance is generally corrective (the failed component is instantaneously replaced by a new one). The notion of rate of occurrence of failure (Rocof) must be introduced, which is generally a time-dependent quantity. However, according to the hypothesis of catastrophic failures, it is known that Rocof is constant and equal to the failure rate λc of catastrophic failures.

For this type of application, repair times are generally negligible compared to the proper operation time so that MTBF ~ MTTF, and therefore MTBF $\approx \frac{1}{\lambda_{\mathrm{obj}}}$.

$$\lambda_{obj} = \lambda_{\mathrm{c}} + \lambda_{\mathrm{HALT}}$$

Resuming the previous hypotheses, it can be reasonably expected that youth and aging failures are not observed, leaving only catastrophic failures so that:

$$\lambda_{\mathrm{target}} = \lambda_{obj} - \lambda_{\mathrm{c}}$$

Therefore, $\lambda_{\mathrm{HALT}} \leq \lambda_{\mathrm{target}}$.

7

Burn-In Test

As already noted, a product that is in the manufacturing stage may typically have youth failures, notably during the first years of operation. Moreover, a product is considered mature provided that it has a constant reliability level that is higher than the specified objective. Prior to being delivered to the system manufacturer, the product should therefore be subjected to a test referred to as a burn-in test (Kececioglu and Sun 1999). The aim of this test is to eliminate as much as possible the defects generated during the manufacturing stage. It is conducted on each of the products delivered, as the manufacturing process may have certain variability and evolve (change of equipment, testing process, operator, variability in the testing means, etc.) in time throughout the period of product delivery.

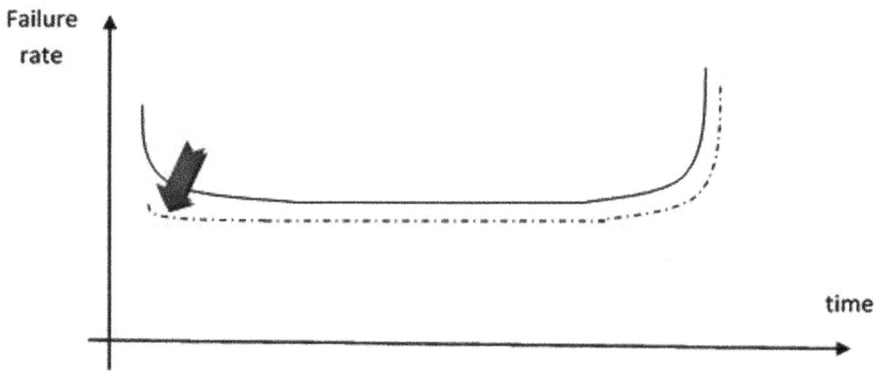

Figure 7.1. *Burn-in test effect on product reliability*

For a color version of all the figures in this chapter, see www.iste.co.uk/bayle/maturity1.zip.

In fact, for the method chosen in this book, this test is part of a more comprehensive method, namely the HALT/HASS method, which involves two distinct stages:

– HALT testing strictly speaking involves finding the operational or environmental limits of the product in terms of thermal and vibration constraints;

– HASS testing involves the burn-in stage.

Burn-in testing presents a challenge, however. The maximum elimination of latent youth defects at reasonable costs (notably the testing duration) requires the application of high physical contributions, significantly higher than those to which the product is subjected during its operation. While this principle may be of interest for the products with latent defects, it is not useful for healthy products (without latent defects). Therefore, it should be verified that the burn-in test does not cause serious damage to healthy products. Moreover, it should also be verified that the burn-in test has certain efficiency (capacity to reveal latent defects), otherwise it would be useless.

Therefore, prior to the actual burn-in test, two tests are conducted on a certain number of representative products of the series:

– POS1 is used to check the minor damage of healthy products;

– POS2 is used to verify the burn-in efficiency.

This is illustrated in Figure 7.2.

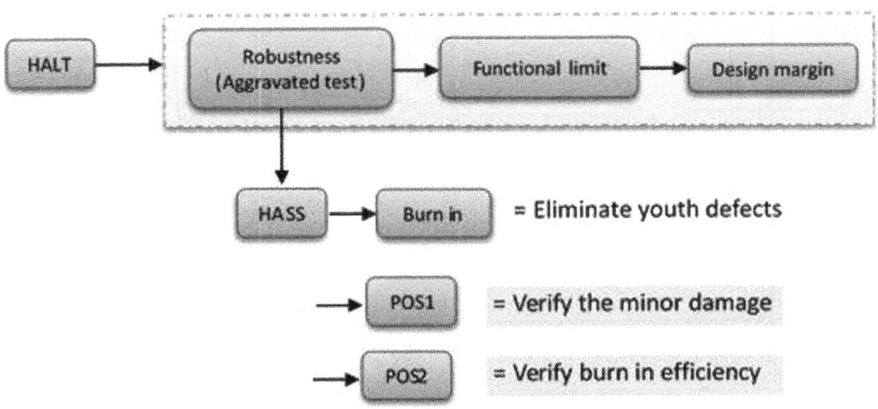

Figure 7.2. *HALT/HASS testing principle*

7.1. Link between HALT and HASS tests

The results of HALT tests, namely the estimation of the operational limits in temperature and vibration, can be used to give initial values for the levels of physical contributions of HASS tests. Any manufacturer willing to conduct a burn-in process must be aware of the following points referring to burn-in:

– there is no universal burn-in;

– burn-in is not a qualifying test. Its objective is not to validate/qualify a batch of products and therefore a technology (with an acceptance criterion such as the maximal acceptable number of defects);

– burn-in does not aim to highlight design defects. This is instead the aim of aggravated and qualifying tests;

– a burn-in operation must lead to breakdowns, otherwise it must be adapted. Two cases are possible:

- if no breakdowns are detected during burn-in and operation, the manufacturing process can be considered as controlled and mature,

- if no breakdowns are detected during burn-in, but there are youth breakdowns in the first years of operation, then the burn-in is not efficient;

– the process should be flexible, fitted to the product.

7.2. POS1 test

POS1 can be used to verify that applying a burn-in cycle in the series stage does not damage the service life of a product by more than "X%". Hence, one or more pieces of equipment are subjected to several cycles. The equipment undergoing this test must be very similar to those to be delivered in the final series. Under no circumstances can this equipment be delivered as operational equipment.

7.2.1. *Miner's approach*

This approach is generally used in HALT/HASS tests to justify the lack of damage of the products without latent defects. The lack of damage rests on a theoretical perspective rooted in Miner's rule (Miner 1945). According to this rule, the cumulative damage of the product is linear. Hence, if one or more products are subjected to "n" burn-in cycles without observing any failures, then the burn-in test cycle to be applied to all the products will damage the healthy products by no more than 1/n. Damage of 2% is generally admitted. In this case, the POS1 test involves the application of 50 burn-in cycles, as illustrated in Figure 7.3.

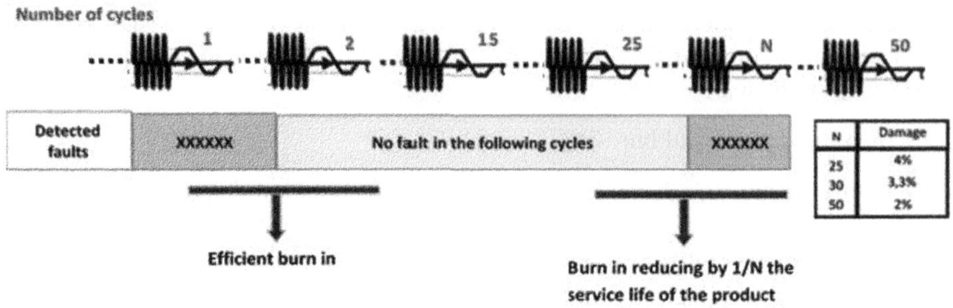

Figure 7.3. *Characteristics of POS1 tests*

A second hypothesis concerns the estimation of damage of products submitted to a HASS cycle. From a reliability perspective, what determines the useful service life of the product (flat sector of the bathtub curve) is the minimum of random variables representing each aging failure mechanism (that can vary depending on the life profile).

This can be schematically represented by Figure 7.4.

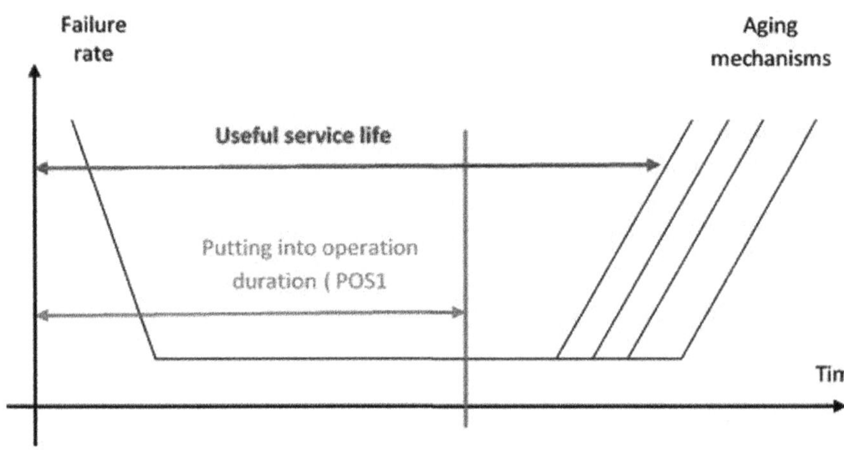

Figure 7.4. *Comparison between service life and putting into service duration on the bathtub curve*

If all the laws of reliability of failure mechanisms were known, it could be verified if the specified reliability objective has been met. Unfortunately, this is not the case, as the component manufacturers do not provide this information in the datasheet of the component. Though experience shows that, in general, aging failure mechanisms are not observed during product operation, the minimum of the aging mechanisms with respect to the product service life is not known. A conservative approach is therefore to take the case in which the minimum corresponds exactly to the end of the service life.

Following the failures observed during the "n" cycles, certain information can be extracted, as listed in Table 7.1.

Cycles									Actions
1	......	15	......	25	......	N	......	50	
	X								Increase constraints
X								X	Proven effectiveness and very robust product
X						X	X	X	Proven effectiveness and robust product
X				X	X	X	X	X	Proven effectiveness, acceptable damage but not robust product
X			X	X	X	X	X	X	Decrease constraints
X		X	X	X	X	X	X	X	Decrease constraints, acceptable damage level decision
X	X	X	X	X	X	X	X	X	No robustness and unfitted burn in

Table 7.1. *Processing of the results of POS1 tests. For a color version of this table, see www.iste.co.uk/bayle/maturity1.zip*

If one or several failures are observed on the first burn-in cycle, this proves a certain effectiveness of burn-in. This scenario is however unlikely given the small number of products undergoing the POS1 test. Hence, POS2 is proposed.

7.2.2. Approach according to the physical laws of failure

There is obviously no physical law of failure taking into account the simultaneous effect of the thermal cycling, vibrations and energizing. Moreover, there is no recent model taking into account the influence of the slope in the thermal cycling, this being particularly important during the precipitation phase. However, from a conservative perspective, it is possible to take into account the aging occurring during the POS1 test by considering only the effect of thermal amplitude according to the Coffin–Manson law. The focus is on the acceleration factor between the POS1 conditions and those observed during operation by the end user. This is given by:

$$AF = \left(\frac{\Delta T_{pos1}}{\Delta T_{opl}}\right)^{m} \quad [7.1]$$

A small problem remains to be solved however, since the thermal amplitude is different between the precipitation and detection phases. However, based on the Sedyakin principle, an equivalent amplitude (Bayle 2019) can be found in terms of reliability, which is given by:

$$\Delta T\_Pos1_{eq} = \left(\frac{\mathrm{N1}.\Delta \mathrm{T_{precipitation}}^{\mathrm{m}} + \mathrm{N2}.\Delta \mathrm{T_{detection}}^{\mathrm{m}}}{\mathrm{N1+N2}}\right)^{1/\mathrm{m}} \quad [7.2]$$

PROOF.–

According to the Sedyakin principle:

$$\int_0^N \left(\frac{\Delta \mathrm{T\_Pos1_{eq}}}{\Delta \mathrm{T_{opl}}}\right)^m .du = \int_0^{\mathrm{N}} \left(\frac{\Delta \mathrm{T_{pos1}(u)}}{\Delta \mathrm{T_{opl}}}\right)^m .du$$

or

$$\int_0^N \Delta \mathrm{T\_Pos1_{eq}}^m .du = \int_0^{\mathrm{N}} \Delta \mathrm{T_{pos1}(u)}^m .du$$

or

$$\Delta \mathrm{T\_Pos1_{eq}}^m .(\mathrm{N1+N2}) = \int_0^{\mathrm{N1}} \Delta \mathrm{T_{precipitation}}^m .du + \int_{\mathrm{N1}}^{\mathrm{N1+N2}} \Delta \mathrm{T_{detection}}^m .du$$

or

$$\Delta T\_Pos1_{eq} = \left(\frac{\mathrm{N1}.\Delta \mathrm{T_{precipitation}}^{\mathrm{m}} + \mathrm{N2}.\Delta \mathrm{T_{detection}}^{\mathrm{m}}}{\mathrm{N1+N2}}\right)^{1/\mathrm{m}}$$

End

Consequently, the acceleration factor can be written as:

$$AF = \left(\frac{\left(\frac{\mathrm{N1}.\Delta \mathrm{T_{precipitation}}^{\mathrm{m}} + \mathrm{N2}.\Delta \mathrm{T_{detection}}^{\mathrm{m}}}{\mathrm{N1+N2}}\right)^{1/\mathrm{m}}}{\Delta \mathrm{T_{opl}}}\right)^{\mathrm{m}}$$

or even

$$AF = \frac{N1.\Delta T_{precipitation}{}^{m} + N2.\Delta T_{detection}{}^{m}}{(N1+N2).\Delta T_{opl}{}^{m}} \qquad [7.3]$$

Therefore, the 50 cycles of POS1 can be reflected by operational cycles in the following form:

$$NB\_Cycles_{opl} = 50.\left[\frac{N1.\Delta T_{precipitation}{}^{m} + N2.\Delta T_{detection}{}^{m}}{(N1+N2).\Delta T_{opl}{}^{m}}\right] \qquad [7.4]$$

EXAMPLE.–

Consider an operational application with the life profile given in Table 7.2.

	ΔT (°C)	Number of cumulated cycles
Brazed joints	20	7,300
Bonding	35	7,300

Table 7.2. *Example of life profile for POS1 test*

Assume the precipitation and detection phases have the following characteristics:

– ΔTprecipitation = 150°C;

– ΔTdetection = 110°C;

– N1 = N2 = 2.

Let us analyze the aging generated by POS1 for the products without defects. For thermal cycling, the following failure mechanisms can be considered:

– Breakage of brazed joints with a coefficient m = 2.

An equivalence in operational cycles is therefore obtained for POS1 for brazed joints

$$NB\_Cycles_{opl} = 50.\left[\frac{2.145^2 + 2.110^2}{4.20^2}\right]$$

or

$$NB\_Cycles_{opl} \approx \ 2\ 163 \text{ cycles}$$

This result is below the 7,300 cycles of the life profile of the previous table. But it is important to note that the acceleration factor model that has been taken into account is very conservative. On the other hand, the focus is on the aging for one burn-in cycle. In this case, it can be seen that it represents 2,163/50 ~ 43 cycles or barely 0.6% (43/7,500).

– Bonding removal or breakage, for active components, with a coefficient m = 4.

$$NB\_Cycles_{opl} = 50.\left[\frac{2.145^4 + 2.110^4}{4.35^4}\right]$$

or

$$NB\_Cycles_{opl} \approx \ 10\,873 \text{ cycles}$$

This result shows that, while being conservative, the POS1 test covers the 7,300 cycles of the life profile considered in this example. If no failures are observed, it is reasonable to believe there will be none during operation.

7.2.3. *Zero-failure reliability proof approach*

It is interesting to see what this means in terms of reliability for the POS1 test. Similar hypotheses to those in section 7.2.2 lead us to obtain:

– Products without maintenance

According to equation:

$$R_{dem} = \exp\left[\frac{\ln(1 - \text{NC}).\left(\frac{\text{Nm}}{\text{AF.NPos1}}\right)^{\beta}}{\text{N}}\right]$$

where:

NC is the level of confidence;

AF is the acceleration factor from equation [7.1];

N is the number of products undergoing POS1 test;

Npos1 is the number of thermal cycles of POS1;

Nm is the number of thermal cycles of the mission;

β is the shape parameter of Weibull law.

Resuming the data in the previous example and moreover considering that N = 4, NC = 90% and β = 5, the following results are obtained:

Brazed joints $R_{dem} = \exp\left[\frac{\ln(1-90\%).\left(\frac{7\,300}{41.4}\right)^5}{4}\right] \approx 72.5\%$

Bonding $R_{dem} = \exp\left[\frac{\ln(1-90\%).\left(\frac{7\,300}{196.4}\right)^5}{4}\right] \approx 99.987\%$

– Products without maintenance

According to the equation:

$$MTBF_{dem} = \text{Npos1. AF.}\,\Gamma\left(1-\frac{1}{\beta}\right).\left(\frac{-N}{\ln(1-NC)}\right)^{1/\beta}$$

Assuming the same hypotheses as previously leads to:

Brazed joints $MTBF_{dem} = (4.50).41.\Gamma\left(1-\frac{1}{5}\right).\left(\frac{-4}{\ln(1-90\%)}\right)^{1/5} \approx$ 8,408 cycles

Bonding $MTBF_{dem} = (4.50).196.\Gamma\left(1-\frac{1}{5}\right).\left(\frac{-4}{\ln(1-90\%)}\right)^{1/5} \approx$ 40,195 cycles

to be compared to 7,300 of life profile.

7.3. POS2 test

As already seen in the previous section, the objective of the POS2 test is to show a certain level of effectiveness of the burn-in cycle. The POS2 cycle relies on four burn-in cycles, and it theoretically damages only 8% of the healthy products according to Miner's criterion. The four cycles are in fact a compromise between reasonable damage and a test duration that is long enough to trap as many latent failures as possible.

This test will be conducted on a series of products that could potentially be delivered to the equipment manufacturer. It is conducted in three successive stages on a certain number of products. As soon as a product is defective, the POS2 test is stopped, irrespective of the stage at which the failure occurs. Conducting the test in three stages makes it possible to limit the number of products undergoing the POS2 test if a failure is detected in stage 1 or 2.

When conducting the POS2 test, several scenarios are possible, as explained in the following table. In this table, we do not quantitatively distinguish the effect of the levels of physical contributions responsible for the occurrence of failures for each cycle. Only a qualitative analysis based on the occurrence of failures can indicate the effectiveness of burn-in testing. In the calculations that follow, the proportion of defect "p" is an unknown quantity. However, depending on the feedback for older products, a relatively accurate estimation of this parameter should be possible. The proposed process is illustrated in Figure 7.5.

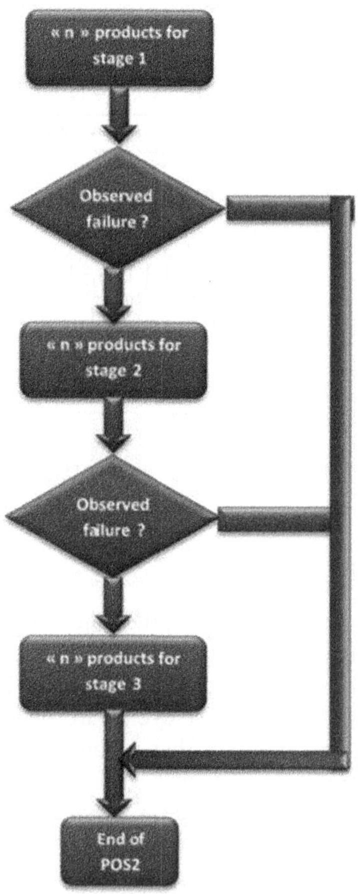

Figure 7.5. *Overview diagram of the POS2 test*

	Stage 1				Stage 2				Stage 3				
	cycle 1	cycle 2	cycle 3	cycle 4	cycle 1	cycle 2	cycle 3	cycle 4	cycle 1	cycle 2	cycle 3	cycle 4	remarks
Scenario 1	0	0	0	0	0	0	0	0	0	0	0	0	Perfect manufacturing process or ineffective burn in
Scenario 2	X												very effective burn in
Scenario 3	0	X											effective burn in
Scenario 4	0		X										fairly effective burn in
Scenario 5	0			X									ineffective burn in
Scenario 6	0	0	0	0	X								very effective burn in
Scenario 7	0	0	0	0	0	X							effective burn in
Scenario 8	0	0	0	0	0	0	X						fairly effective burn in
Scenario 9	0	0	0	0	0	0	0	X					ineffective burn in
Scenario 10	0	0	0	0	0	0	0	0	X				very effective burn in
Scenario 11	0	0	0	0	0	0	0	0	0	X			effective burn in
Scenario 12	0	0	0	0	0	0	0	0	0	0	X		fairly effective burn in
Scenario 13	0	0	0	0	0	0	0	0	0	0	0	X	ineffective burn in

Table 7.3. *Possible scenarios during POS2 testing. For a color version of this table, see www.iste.co.uk/bayle/maturity1.zip*

All the possible scenarios are defined in Table 7.3.

POS2 testing is conducted without transfer of products from one stage to the next. Hence, it is possible to model the POS2 test by hypergeometric distribution. Hence:

– Q is the number of products undergoing burn-in (one HASS cycle);

– n is the number of products undergoing POS2 testing;

– p is the proportion of products with a latent defect.

The probability of observing "k" failures is given by:

$$Pr(n,k,p,Q) = \frac{C_{p.Q}^{k} . C_{(1-p).Q}^{n-k}}{C_{Q}^{n}} \quad [7.5]$$

What we are interested in here is estimating the probability of having at least one failure for POS2 in order to verify the burn-in effectiveness. This probability can be written as:

$$Pr(n,p,Q) = 1 - P(n,0,p,Q)$$

or:

$$Pr(n,p,Q) = 1 - \frac{C_{(1-p).Q}^{n}}{C_{Q}^{n}} \quad [7.6]$$

7.3.1. *Influence of parameter Q*

Let us fix p = 2% and observe the evolution of this probability as a function of Q and n. Figure 7.6 is obtained.

It can be seen that the number of products Q subjected to burn-in does not influence the probability of detecting a defect during POS2 testing. Furthermore, it is important to acknowledge that it is necessary to have a significant number of products for POS2 testing in order to achieve a reasonable probability of detecting one or more latent defects.

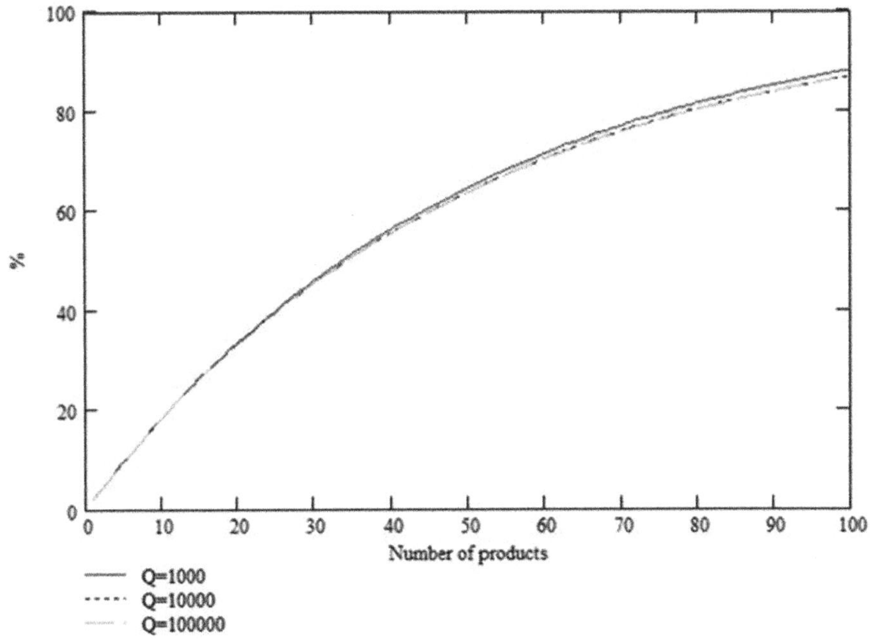

Figure 7.6. *Probability to detect at least one failure influenced by Q*

7.3.2. *Influence of parameter p*

Let us fix Q = 1,000 and observe the evolution of this probability as a function of p and n. Figure 7.7 is obtained.

It can be noted that the proportion of defects has a significant impact on the probability of detecting at least one defect during the POS2 test. Moreover, in order to have a reasonable probability of detecting a defect, the required number of products is all the larger as the defect proportion is larger, which is in fact logical.

Indeed, since generally p << 1, equation [7.6] can be approximated by:

$$f(n) \approx 1 - \exp(-\mathrm{p.n}) \quad [7.7]$$

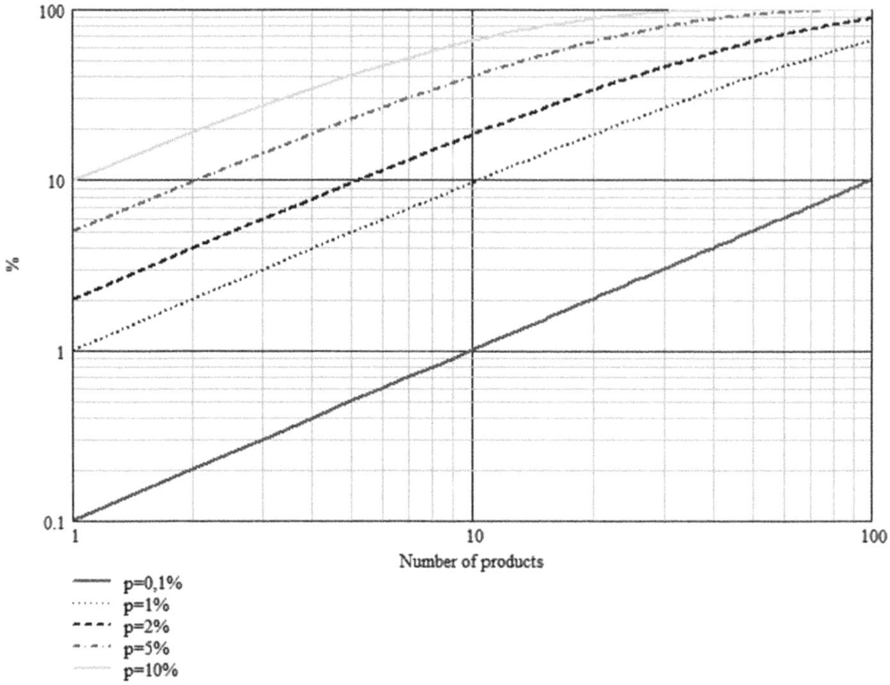

Figure 7.7. *Probability to detect at least one failure influenced by p*

Hence, the value of n corresponds to a given probability

$$n \approx -\frac{\ln(1-\text{Pr})}{\text{p}} \qquad [7.8]$$

Based on equation [7.8], it is possible to set a probability to filter a defect and estimate the number of products "n" that must be subjected to POS2 testing. But how can this probability be fixed knowing that it is an ascending function of n? The economic aspect can then open up a possibility. Indeed, given:

– Ce is the cost of POS2 test;

– Cp is the cost of a product;

– Cr is the cost of repair due to manufacturing defect;

– Ct is the total cost,

it can be written that:

$$Ct = \mathrm{Ce} + \mathrm{n.Cp} + \big(1 - \mathrm{Pr(n)}\big).\mathrm{Q.p.Cr}$$

Indeed, the cost due to products having a defect depends on their number Q.p as well as on the fact that they were not filtered out by burn-in. Based on equation [7.8], it can be written that:

$$Ct \approx \mathrm{Ce} + \mathrm{n.Cp} + \exp(-\mathrm{p.n}).\mathrm{Q.p.Cr} \qquad [7.9]$$

This equation is differentiated with respect to n:

$$\frac{\delta Ct}{\delta n} = \mathrm{Cp} + \exp(-\mathrm{p.n}).\mathrm{Q.p.Cr}.(-\mathrm{p})$$

This derivative is zero if:

$$\mathrm{Cp} + \exp(-\mathrm{p.n}).\mathrm{Q.p.Cr}.(-\mathrm{p}) = 0$$

or

$$(-\mathrm{p.n}) = \ln\left(\frac{\mathrm{Cp}}{\mathrm{Q.p^2.Cr}}\right)$$

or finally

$$\mathrm{n} = \frac{\ln\left(\frac{\mathrm{Q.p^2.Cr}}{\mathrm{Cp}}\right)}{\mathrm{p}} \qquad [7.10]$$

Since "n" is a non-zero integer, the following inequality must be valid: $\frac{\mathrm{Q.p^2.Cr}}{\mathrm{Cp}} > 1$.

NOTE.–

– The cost of the POS2 test is not involved in the search for the optimum number of products undergoing POS2 test.

– Due to the previous inequality, it is possible to have no optimum.

– In case there is one, the optimum is a minimum since

$$\frac{\delta^2 Ct}{\delta n^2} = \exp(-p.n).Q.p^3.Cr > 0$$

– The minimal cost obtained is given by:

$$Ct_{Min} = Ce + \frac{Cp}{p}.\left[\ln\left(\frac{Q.p^2.Cr}{Cp}\right) + 1\right]$$

EXAMPLE.–

Ce = 20,000

Cp = 200

Cr = 10,000

Q = 10,000

P = 2%

The resulting optimal number of products for POS2 test is n* = 150. The total corresponding cost is ~ 420,000.

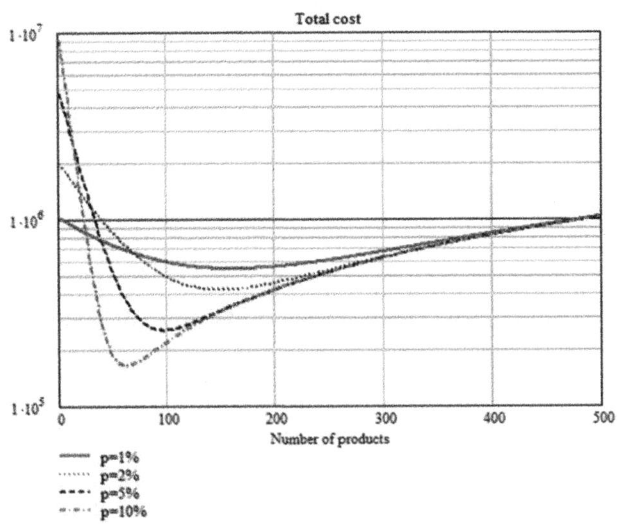

Figure 7.8. *Example of cost optimization taking burn-in into account*

7.3.3. *Summary of the POS2 test*

It should be recalled that the objective of the POS2 test is to estimate the effectiveness of burn-in. The number of products subjected to burn-in is set in advance, and therefore this parameter cannot be changed. Neither can the proportion "p" of defects; it simply remains the number of products at each POS2 stage to optimize the probability of having a failure in this test.

If no failure is observed after the third stage of POS2 testing, several hypotheses can be formulated to explain this fact:

– The manufacturing process is excellent (p = 0.1%), and few defects are generated during this stage. The previous figure shows that with 100 products having undergone POS2 testing, there is only 10% chance of observing a failure. No conclusion can be drawn on burn-in, except that it may be useless. This situation must be confirmed by the absence of youth failures with the final user.

– Burn-in is not at all effective and does not filter out any latent defects. Theoretically, there are 2 solutions:

- wait for feedback information that confirms whether or not there are youth failures observed by the end user;

- or, modify the levels of physical contributions implemented for burn-in, as they are not considered serious enough.

The general rule is to wait for the operational feedback information.

7.4. HASS cycle

Once the POS1 and POS2 tests are conducted, the HASS cycle involves two distinct stages.

7.4.1. *Precipitation stage*

The precipitation stage is conducted with the equipment in operation and under continuous monitoring. However, there is no need to stop the test during the precipitation stage if there are functional drifts. As a general rule, the temperature limits, HOL and LOL, during the precipitation stage are defined using the results of HALT with a 10°C reduction applied to the HOT and LOT limits:

– high operational limit: HOL = UTOL – 10°C;

– low operational limit: LOL = LTOL + 10°C.

A slightly different definition can sometimes be found in which the temperature limits are defined based on HALT results with a temperature reduction applied to the UTOL and LTOL limits of the HALT test defined as follows:

– high operational limit: HOL = HOT – 10% x HOT;

– low operational limit: LOL = LOT + 10% x LOT.

In the case of thermal cycling (type RCT), the temperature ramps are fixed at the value found during the tests for the VRT limit search; this can go up to 60°C/min in the chamber per piece of equipment (10°C/min locally on the burnt-in material). These temperature ramps are instructions of the HALT/HASS chambers (QUALMARK type). It is important to instrument any equipment confined without forced ventilation. In general, the maximum value reached is 35 to 45°C/min at the product level.

The vibration limit HV during the precipitation stage is determined from the HALT test and is decreased by 50%: HV = HVOL/2. It is important to instrument the equipment when several pieces of equipment are in the chamber in order to verify the homogeneity of vibration levels after installations.

7.4.2. *Detection stage*

In the detection stage, the constraints are generally applied at the specification level, making it possible to evidence obvious defects. The detection stage is conducted with the equipment under operation and under monitoring. The high and low limits of temperature during the detection stage rely on the maximal operational specification levels, and there is no notion of margin.

The temperature ramps can be weak, for example, 3°C or 10°C per minute if the equipment is to be tested during these ramps. The higher limit of vibration is reduced by 60 to 80% of the value fixed in precipitation, with the definition of a minimum level generally of 5 Grms, below which there is no descent.

The applied vibration can have a sawtooth form, ascending and descending by 5 Grms to X% of the value fixed in precipitation or in a constant manner.

At the end of each level, there is an on/off stage that may involve the rated voltages and currents as well as, in certain cases, with the minimum, maximum levels of voltage and current.

Figure 7.9 proposes an example of the HASS cycle that combines temperature levels and random vibrations.

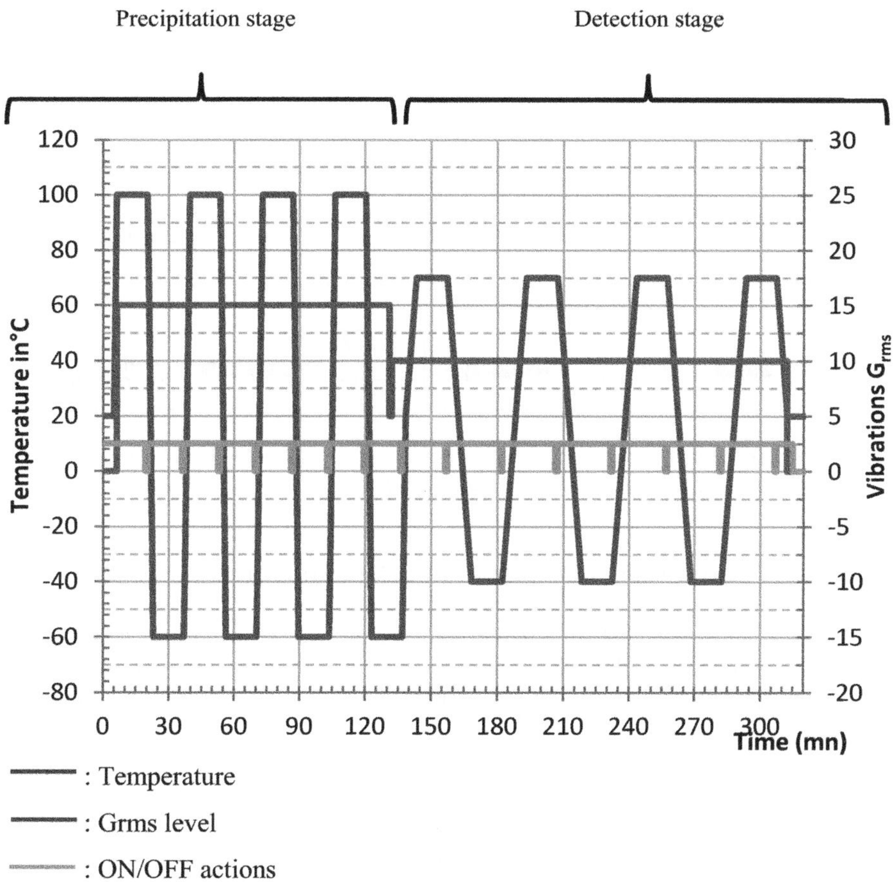

Figure 7.9. *Example of burn-in cycle according to HALT/HASS method*

The overall HALT/HASS method is illustrated in Figure 7.10.

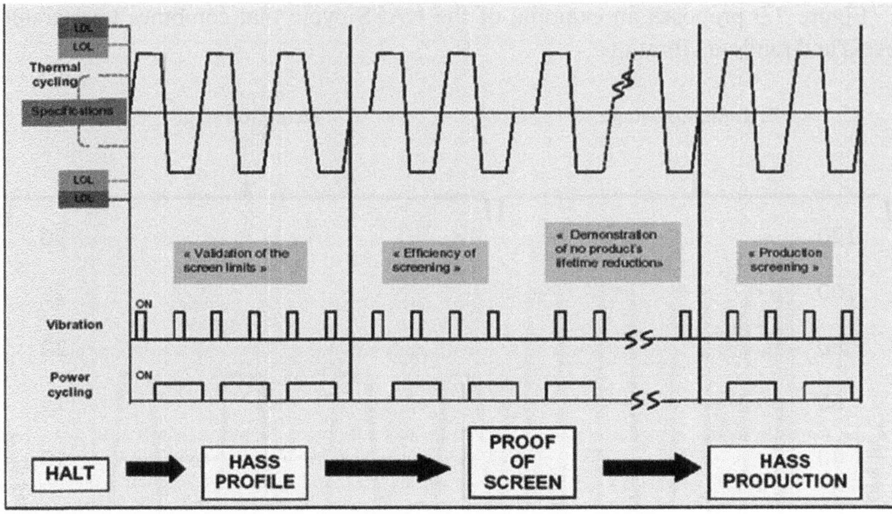

Figure 7.10. *Illustration of the stages of the HALT/HASS method. For a color version of this figure, see www.iste.co.uk/bayle/maturity1.zip*

7.5. Should burn-in tests be systematically conducted?

At first glance, the cost of a manufacturing process with burn-in seems higher than that of a process without burn-in. The overall cost must nevertheless integrate the costs induced by youth defects, which makes it possible to highlight the economic interest of burn-in.

We should therefore ask the question whether implementing a burn-in process is cost-efficient with respect to the objectives set. Implementing a burn-in process may be wise if the manufacturer expects to improve the following points:

– monitoring the possible drift of manufacturing;

– decrease of the number of refused acceptance tests;

– reduction of the number of failures at the expense of the manufacturer;

– reduction of the logistics due to the decrease in the number of failures;

– preservation/improvement of the brand image.

Economic criteria are completed by certain manufacturing constraints that can guide the choice of conducting a burn-in test or not.

7.5.1. *Constraints extrinsic to the equipment manufacturer*

Several constraints extrinsic to the equipment manufacturer can be considered:

– recommendation of the system;

– operational context (life profile, production volume, etc.);

– a diminished brand image in front of the client;

– the seriousness of the operational life profile;

– existence of strong competition.

7.5.2. *Constraints intrinsic to the equipment manufacturer*

Several constraints intrinsic to the equipment manufacturer can be considered:

– recommendation by the company strategy;

– cost constraints in the case of poor reliability of the product being operated by the client;

– very demanding reliability objectives;

– the quantity of products to be delivered.

Hence, the economic criteria and the manufacturing constraints guide the choice for or against a burn-in process.

7.5.3. *Decision criteria*

The overview diagram given in Figure 7.11 proposes a certain number of qualitative manufacturing criteria that may influence the decision of the equipment manufacturer to implement a burn-in process. Other criteria and constraints may be involved in the final decision (see below).

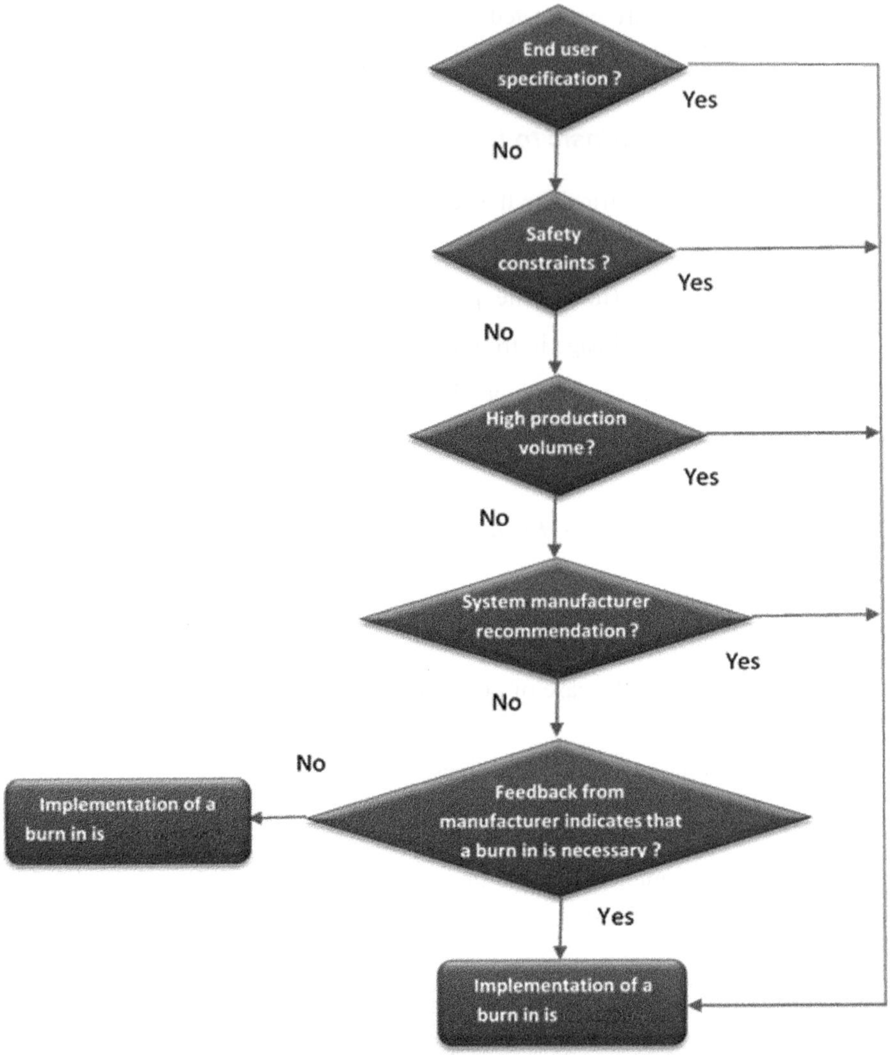

Figure 7.11. *Overview diagram of decision-making*

– Final user specification:

The final user can, according to their own experience, require a burn-in test to be conducted and include it in the specifications they send to the manufacturer. In this case, it is a contractual demand that states the necessity for the manufacturer to conduct a burn-in test. This situation is typically encountered in the case of space equipment, in applying the ECSS standards required by the users. If not specified by the final user, the choice in conducting a burn-in test is made according to the other criteria given in Figure 7.11.

– Safety constraints:

If the manufactured equipment/product is such that:

- a high reliability level of the full product is required (typically civilian avionics);

- if it is defective due to a youth failure, it endangers the operation of the complete product, a burn-in process should be implemented, which will eliminate a maximum number of youth defects in order to reach the required reliability level.

– Production volume:

If the manufacturer has the elements described in section 7.5.3 available, it is possible to calculate if implementing a burn-in process is of interest, depending on the number of parts to be manufactured. Indeed, depending on the coverage rate, the effectiveness of the burn-in process and the proportion of products with latent defects, the minimal quantity to be produced before detecting a defect may be higher than the overall production. In this latter case, implementing a burn-in process presents no interest. If the manufacturer does not have sufficient elements, the choice of conducting a burn-in process is made according to the other criteria in Figure 7.2.

– System equipment recommendation:

If the equipment/manufactured product is integrated by a system manufacturer before delivery to the final client, it can, depending on its own experience, or in response to client demand, require the implementation of a burn-in and include it in the specifications it sends to the manufacturer.

– Feedback from the manufacturer:

In the absence of demands or external elements leading to the implementation of a burn-in process, the feedback from the manufacturer on similar products/equipment can be used to guide its choice. If for a product/equipment of similar complexity, involving identical manufacturing processes, burn-in implementation has previously revealed a level of latent defects that was unacceptable with respect to the reliability level expected for the new product, a burn-in process would logically be once again implemented.

– Interest of burn-in implementation:

This analysis can be conducted when parameters such as coverage rate, effectiveness and proportion of products with latent defects are known or estimated.

The precondition for burn-in implementation is to have the possibility to observe products with youth defects during burn-in.

Several parameters are involved in this case:

– the volume N (quantity) of products to be delivered. This is a purely deterministic quantity known in advance;

– the coverage rate τc of the monitoring implemented during burn-in. This is the capacity of burn-in to be able to observe a product failure. This quantity is estimated;

– burn-in effectiveness τeff, meaning its capacity to transform a latent failure into an obvious one;

– the proportion p of products having a latent failure induced by the manufacturing process. This quantity is difficult to estimate precisely, as it may vary throughout the manufacturing process (variability of means, operators, components, etc.).

The number of products Q for which a manufacturing defect can be revealed can then be estimated by:

$$Q = \mathrm{p.N.}\tau_{\mathrm{c}}.\tau_{\mathrm{eff}} \quad [7.11]$$

The proportion of products Pr for which a manufacturing defect is observed is then:

$$Pr = \frac{\mathrm{Q}}{\mathrm{N}} = \mathrm{p.}\tau_{\mathrm{c}}.\tau_{\mathrm{eff}} \quad [7.12]$$

For the probability of observing defects revealed by burn-in to be as high as possible, it is our proposal to estimate the lower bound of this proportion Pr for a given risk level α (Gau 2010).

This is given by:

$$\mathrm{Pr}_{\mathrm{Inf}} = \frac{1}{1+\left(\frac{N-Q+1}{Q}\right).qF\left(1-\frac{\alpha}{2};2.(N-Q+1);2.Q\right)} \quad [7.13]$$

where qF is the quantile of Fisher–Snedecor distribution (Gau 2010).

The quantile of a distribution law is a reciprocal function of the corresponding distribution function. The quantity of the Fisher–Snedecor law can be easily calculated, for example, using Excel with the function “INVERSE.LOI.F”.

Hence, if N = 100, α = 10% and Q = 2, the proportion of defects amounts to 2%, but the lower bound at risk α = 10% amounts to 0.36%. The lower bound of the number of products having a youth defect revealed by the burn-in can now be estimated by burn-in from the following relation:

$$Q_{Inf} = \frac{N.\tau_c.\tau_{eff}}{1+\left(\frac{N-N.p+1}{N.p}\right).qF\left(1-\frac{\alpha}{2};2.(N-N.p+1);2.N.p\right)} \quad [7.14]$$

EXAMPLE.–

Given the following data:

– 20 < N < 100,000

– τ_c = 90%

– τ_{eff} = 80%

– 1% < p < 10%

The curve shown in Figure 7.12 is obtained.

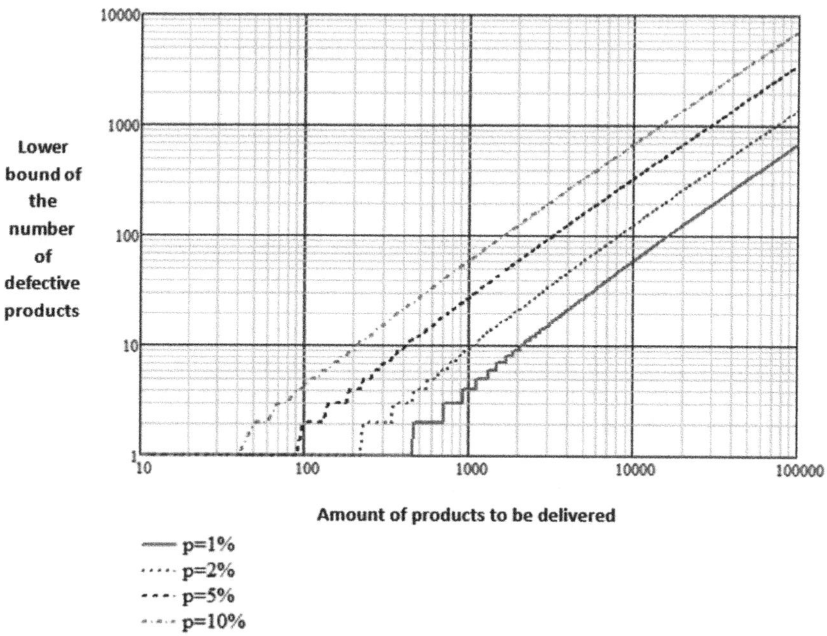

Figure 7.12. *Lower bound of the number of defective products*

Hence, estimating that p = 5%, a burn-in presents interest provided that the number of defective products is above 100. Moreover, for a total of 100,000 products installed, even with a defect proportion of 1%, 677 defective products could be detected over the 1,000 products generated by the manufacturing process.

Obviously, while the amount of products to be delivered is known, the proportion of defective products cannot be known in advance. However, a reasonable estimation is possible due to the experience with products from previous generations. Setting a minimum of 10 potentially defective products, the curve shows that the least number to be delivered are:

– 200 products if the proportion of defects is 10%;

– 500 products if the proportion of defects is 5%;

– more than 1,000 products if the proportion of defects is 2%;

– more than 2,000 products if the proportion of defects is 1%.

7.6. Test coverage

Throughout the manufacturing stage, a certain number of tests with various objectives are conducted. It is important to know the coverage of these tests, meaning the capacity of a test to detect manufacturing defects.

This is referred to as the test coverage rate.

Indeed, it is important to be able to precipitate the latent failures into obvious failures, but it should be possible to observe the latter, hence the importance of the test coverage.

In general, the client requirements specify a certain coverage rate of the tests, but only during the operation stage. Nothing is specified during the manufacturing stage and, strictly speaking, it is up to the equipment manufacturer to propose certain demands on this subject.

This is not always (or often) done on each of the tests implemented during the manufacturing stage, and there is no dedicated generic methodology.

According to the general definition of coverage rate of a test, the following definitions are applicable:

– Coverage rate of the manufacturing process

Ratio between the number of defects detected by the test and the number of potential defects. Here, the term defect refers to a defect induced by the manufacturing process (improper component, missing component, improper value of the component, etc.).

– Coverage rate of the burn-in test

Ratio between the sum of failure rates of the failure modes detected by the burn-in process and the failure rate of the product.

– Coverage rate of the acceptance test

Ratio between the failure rate of the failure modes detected by the acceptance test and the failure rate of the product.

NOTE.– The various failure rates result from analyses of predictive reliability.

– Industrial application

Indeed, a space application (such as Ariane rocket) does not have the same level of demand as a "consumer" application (e.g. mobile telephone).

– Manufacturing stage

Each manufacturing stage has a specific objective, and the demand and the method for estimating the coverage rate is different.

Nowadays, the notion of coverage rate is present for the receipt test but not always for the other stages of the manufacturing process. The coverage rate results from the FMEA analysis for the qualitative part and from the predictive reliability analyses for the quantitative part.

The following definitions are proposed:

- Potential failures: any type of failure concerning the components of the tested product.

- Failures without effect: any type of failure that has no operational effect on the product.

- Detected failures: any type of failure that could be detected by the implemented test.

The coverage rate of a test is generally defined by:

$$\tau_1 = \frac{\lambda_{detected\_failures}}{\lambda_{potential\_failures}}$$

However, this definition does not seem to provide enough information, and it is preferable to associate it with the testability rate of the test, expressing the ability of a product to be tested. Moreover, a definition of the effectiveness rate of a test is proposed, which characterizes its capacity to detect detectable breakdowns, hence the following definition:

$$\tau_{\text{Effectiveness}} = \frac{\lambda_{detected\_failures}}{\lambda_{potential\_failures - \lambda_{failures\ without\ effect}}}$$

This definition is often used in the notion of coverage rate of a test.

7.7. Economic aspect of burn-in

A proportion "p" of products is considered to have youth defects. Consider an overall cost C(t) comprising the following:

– fixed costs of burn-in;

– hourly costs of burn-in;

– cost of repair of the defective products subjected to burn-in;

– cost of repair of defective products during operation.

The following notations are made:

– CSP: fixed costs of burn-in;

– CTP: hourly costs of burn-in;

– CHP: cost of repair of the defective products subjected to burn-in;

– CFS: cost of repair of defective products during operation;

– Tb: burn-in duration;

– Texp: operational duration;

– τc: burn-in coverage rate;

– N: number of products to be delivered;

– fw: density of probability of a product having a youth failure, modeled by a Weibull distribution of parameters η and β;

– Fw: probability of failure of a product having a youth failure according to the Weibull law;

– fe: density of probability of a product that has no youth defect, modeled by an exponential distribution of parameter 1/MTBF;

– p: proportion of products having a youth defect;

– $f^{<i>}$: convolution product "i times by itself" of the density of probability f;

– N1(t): average number of breakdowns of the products with youth defects;

– N2(t): average number of breakdowns of the product during operation.

7.7.1. *No burn-in test is conducted*

The overall cost induced by the products depends on:

– the repair cost C1 during operation of products having youth defects or:

$C_1(\mathrm{t}) = \mathrm{N.C_{FS}.p.N1(t)};$

– the repair cost C2 during operation of products having no youth defects or:

$\mathrm{N.C_{FS}.(1-p).N2(t)}.$

The overall cost is therefore given by:

$$C_{SD}(\mathrm{t}) = \underbrace{\mathrm{N.p.C_{FS}.N1(t)}}_{\text{Products with defects}} + \underbrace{\mathrm{N.C_{FS}.(1-p).N2(t)}}_{\text{Products without defects}}$$

or even:

$$C_{SD}(\mathrm{t}) = \mathrm{N.C_{FS}.\left(p.N1(t) + (1-p).N2(t)\right)} \qquad [7.16]$$

The average number of breakdowns N2(t) during operation for the products without manufacturing defects can be calculated using the Poisson process theory. Indeed, the observed catastrophic breakdowns can be modeled by an exponential distribution of parameter λ. Considering the effect of maintenance, a homogeneous Poisson process is obtained, whose average number of breakdowns at instant "t" is given by:

$$N2(t) = \lambda.t = \frac{t}{MTBF}$$

The average number of breakdowns N1(t) during operation for the products with manufacturing defects can be calculated using the renewal process theory (Bayle 2019) and is given by:

$$N1(t) = \int_0^t \sum_{i=1}^{+\infty} \left[fw(u) * fe^{<i-1>}(u) \right].du$$

1st breakdown : youth failure

The following breakdowns are catastrophic after the repair of youth defects

This can be illustrated with the following diagram:

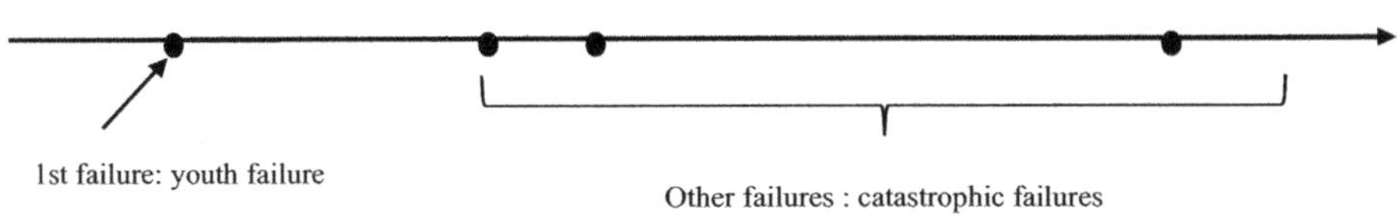

7.7.2. *Burn-in test is conducted*

As already noted, the set of products can be divided into two categories:

– those that have youth defects, and their number is p.N;

– those that have no youth defects (healthy products), and their number is (1 - p).N.

Among the products having a latent youth defect, a certain number will be detected, while others will not, because the coverage rate of the tests is generally not 100%. Figure 7.13 illustrates the various possible cases for a product.

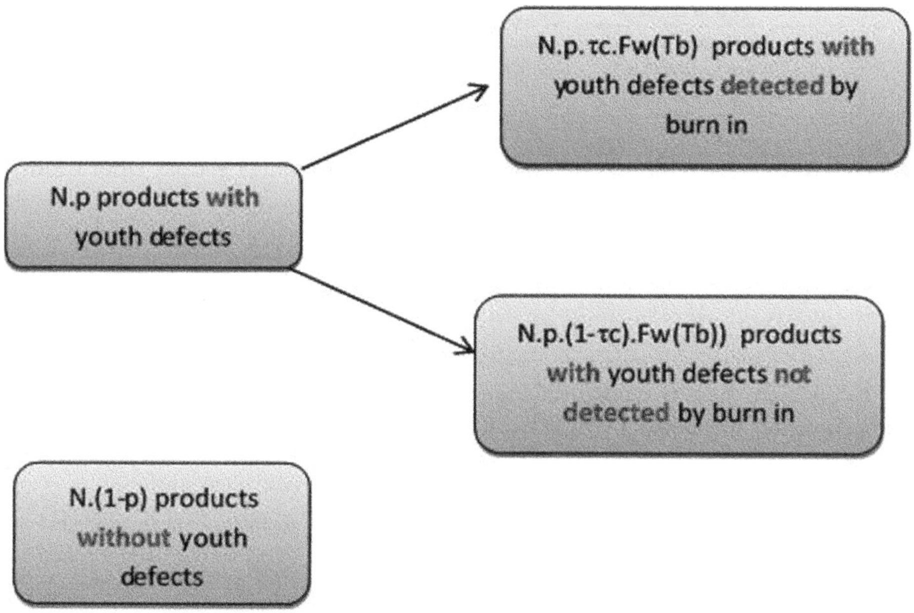

Figure 7.13. *Possible scenarios for the distribution of products with or without defects*

The overall cost C(t) induced by the products depends on:

– burn-in cost given by:

$$\mathrm{Cost}_{\mathrm{Burn}} - \mathrm{in}(t) = \underbrace{\mathrm{CSP.N}}_{\text{Fixed cost}} + \underbrace{\mathrm{CTP.N.Tb}}_{\text{Hourly cost}} + \underbrace{\mathrm{p.\tau c.Fw(Tb).CHP.N}}_{\text{Cost of repair of youth defects } \textbf{filtered out by burn-in}};$$

– operation cost given by:

$$\text{Cost}_{\text{Operation}}(t) = N.C_{FS}.(1-p).N2(t) + N.C_{FS}.p.\tau c.Fw(Tb).N2(t) + N.p.(1-\tau c.Fw(Tb)).C_{FS}.N1(t)$$

Cost of repair of **catastrophic failures of** products **without** youth defects

Cost of repair of **catastrophic failures** of products **with** youth failures **filtered out by burn-in**

Cost of repair of products **with** youth failures **not filtered out by burn-in**

The overall cost with burn-in is: $C_{AD}(t) = \text{Cost\_Burnin(t)} + \text{Cost\_Operation(t)}$ or still:

$$C_{AD}(t) = N.C_{FS} + N.C_{TP}.T_b + N.p.\tau c.Fw(T_b).CHP + N.C_{FS}.p.\tau c.Fw(T_b).\big(N2(t) - N1(t)\big) + C_{SD}(t)$$

To compare these two costs, it is important to study the difference between them. Let us note this difference by $\Delta C(t) = C_{AD}(t) - C_{SD}(t)$ or:

$$\Delta C(t) = N.C_{FS} + N.C_{TP}.T_b + N.p.\tau c.Fw(T_b).C_{HP} + N.C_{FS}.p.\tau c.Fw(T_b).\big(N2(t) - N1(t)\big).$$

The calculation of N1 is complex and requires knowledge on the parameters of the Weibull law modeling youth breakdowns. To simplify the theoretical approach, the following approximation can be made (as they are differentiated only by a youth failure):

$$N1(t) - N2(t) = 1$$

Hence:

$$\Delta C\ (t) \sim N.\big(C_{SP} + C_{TP}.T_b + p.\tau c.Fw(T_b).(C_{HP} - C_{FS})\big)$$

Numerical applications

– η = 1 hour;

– β = 0.7;

– C_{TP} = 0.2 k€;

– C_{SP} = 2 k€;

– C_{HP} = 2 k€;

– C_{FS} = 10 k€;

– τc = 90%;

– MTBF = 1,000 hours;

– N = 100;

– Texp = 10,000 hours.

EXAMPLE 7.1.– Poor manufacturing process and effective burn-in:

– p = 10%;

– Tb = 10 hours.

The curve shown in Figure 7.14 is obtained.

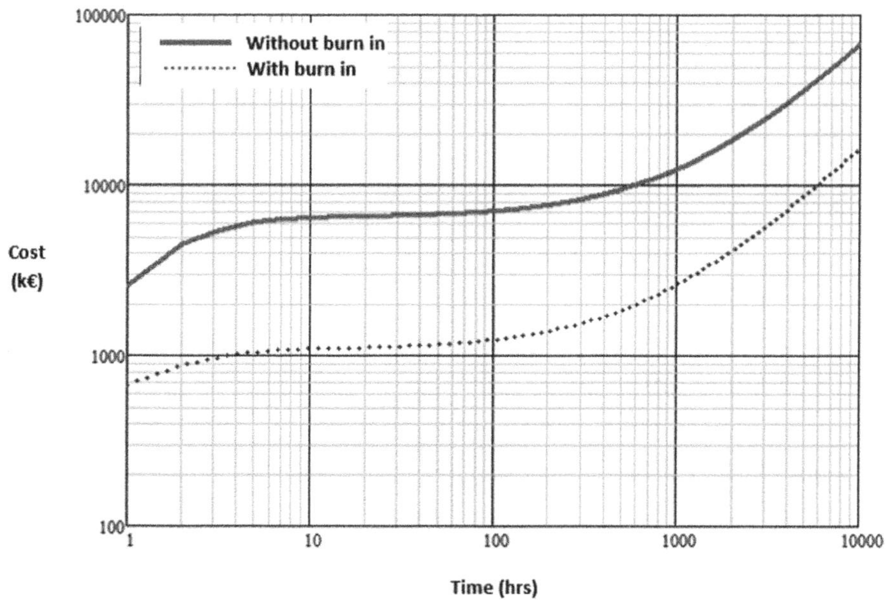

Figure 7.14. *Cost of manufacturing process depending on time*

The cost with burn-in is lower than that without it. This is a logical result, since there is a poor manufacturing process (p = 10%) and an effective burn-in (τc = 90% and Tb = 10 hrs). Indeed, due to the effectiveness of burn-in, most of the youth failures are filtered out and they will not occur during operation. Since the repair cost during operation is higher than that of burn-in, the overall result is positive.

EXAMPLE 7.2.– Very good manufacturing process and ineffective burn-in:

– p = 0.1%;

– Tb = 0.1 hrs.

The curve shown in Figure 7.15 is obtained.

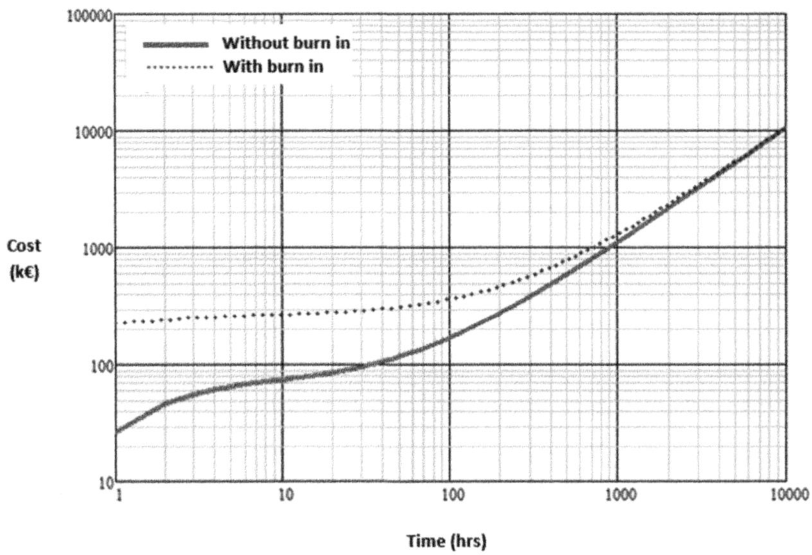

Figure 7.15. *Cost of manufacturing process depending on time – example 1*

The costs with burn-in are higher than that without it. Indeed, though burn-in may be effective (Tb = 0.1 hrs and τc = 90%), since the manufacturing process is very good (p = 0.1%), burn-in filters out few youth failures and is therefore more expensive.

EXAMPLE 7.3.– Very good manufacturing process and effective burn-in:

– p = 0.1%;

– Tb = 10 hours.

The curve shown in Figure 7.16 is obtained.

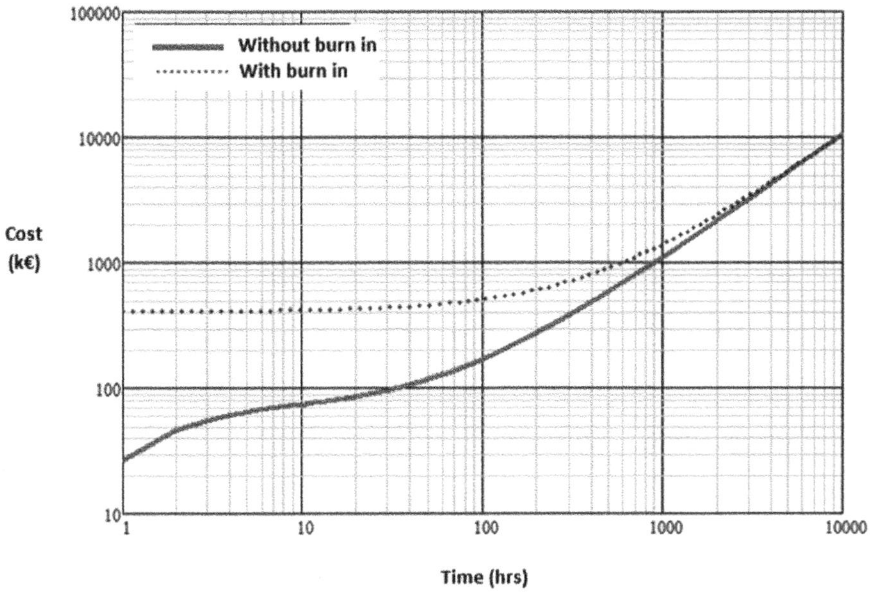

Figure 7.16. *Cost of the manufacturing process depending on time – example 2*

In this example, the overall cost with burn-in is higher than that without it. This is a logical result, since the manufacturing process is very good (p = 0.1%) and therefore generates only a few youth failures during operation. Hence, burn-in adds an additional cost without return on investment.

EXAMPLE 7.4.– Poor manufacturing process and ineffective burn-in:

– p = 10%;

– Tb = 0.1 hrs.

The curve shown in Figure 7.17 is obtained.

The costs with and without burn-in are pretty much identical. Indeed, though the manufacturing process generates a proportion of products with latent failures (p = 0.1%), burn-in being ineffective, few youth failures are filtered during burn-in.

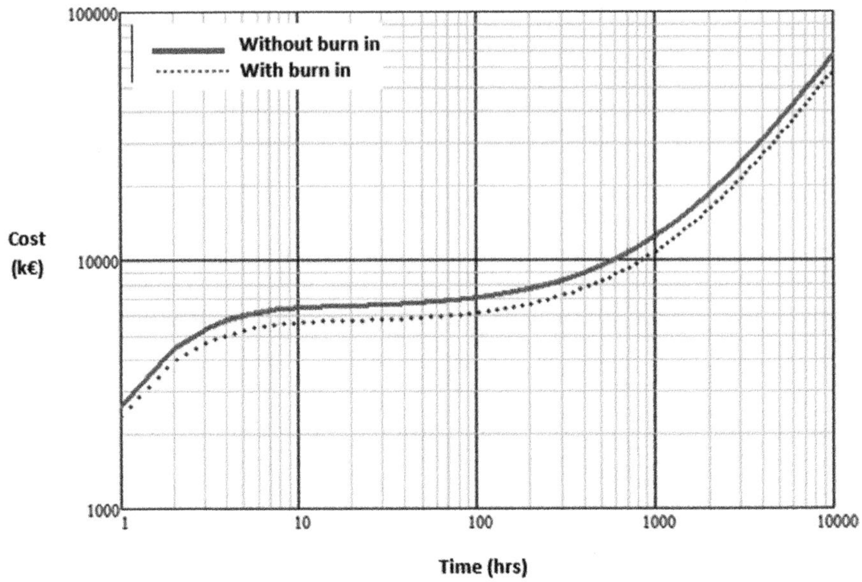

Figure 7.17. *Cost of manufacturing process depending on time – example 3*

8

Run-In

8.1. Run-in principle

A particular type of burn-in is known as the run-in. This concept was well-known several decades ago, notably for automobiles. In the past, the purchase of a new car was accompanied by a run-in stage: longer or shorter, not too rapid, not too high in rpms, daily check-up of levels, etc.

Figure 8.1. *Run-in of a car engine. For a color version of this figure, see www.iste.co.uk/bayle/maturity1.zip*

First, let us recall what the run-in of a new car entailed. It covered a period of time between the purchase and the first few hundred kilometers traveled, say up to 1,000 km on the counter. The idea was to drive with almost excessive caution, allowing all of the parts and gears to properly interconnect and to erase all machining traces. Conversely, being impatient and pushing the gas pedal to the floor could generate even more friction between the parts, distorting them and increasing oil consumption and, consequently, the number of failures.

On the contrary, a failure is not always defined as the cessation of a function of a component, or its destruction. Indeed, it can also be about no longer reaching a certain level of performance. This is particularly the case with components that have a "reference" function, for example a voltage regulator (electrical voltage), an oscillator (frequency), etc.

These components generally provide a physical quantity with a certain level of precision guaranteed by the manufacturer. However, this level of precision is not always sufficient to meet the requirements of a product specification. Moreover, component performance most often degrades over time and as a result of the temperatures that it is exposed to.

As an illustration, consider the example of a logarithmic drift of equation:

$$D(t) = \mathrm{a}.\ln(\mathrm{t}) + \mathrm{b} \qquad [8.1]$$

with a = 1.1 and b = -2.4.

This is illustrated in Figure 8.2.

It can be seen that the drift of the component performance almost reaches a value of 5.2 at time = 1,000 hours. Assume that the required specification is 3. As such, this component is not suitable for guaranteeing the specification, although the manufacturer guarantees the best performance.

The idea is then to submit the component to a test, referred to as a stabilization test, at the equipment manufacturer before delivering it to the system manufacturer. Indeed, when calculating the slope of this drift, namely the drift rate in time, it is obvious that it decreases, being inversely proportional with time. In other terms, the component drifts a lot at the beginning and less and less later on. Hence, if the component is submitted to a sufficiently long test at the equipment manufacturer, the resulting component has a drift at the system that meets the demand of the specification.

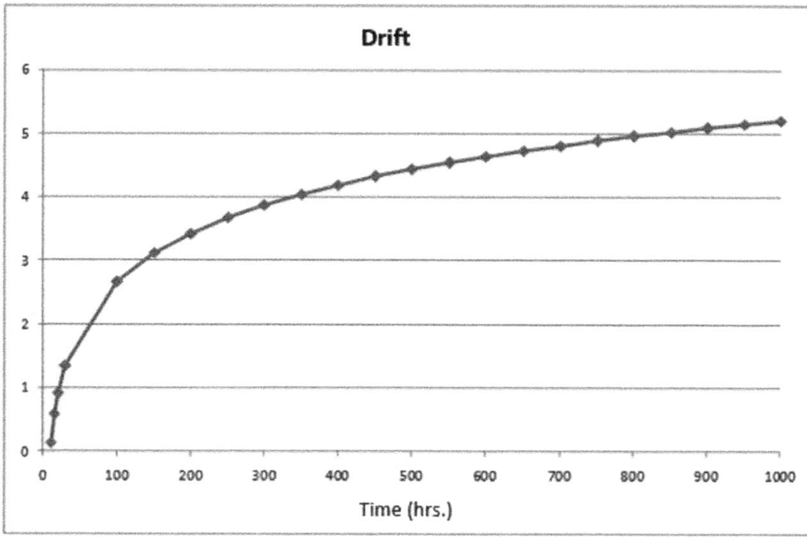

Figure 8.2. *Example of concave degradation*

Assume that the component is subjected to a test (referred to as stabilization test) of 100 hours. Based on the previous example, Figure 8.3 is obtained.

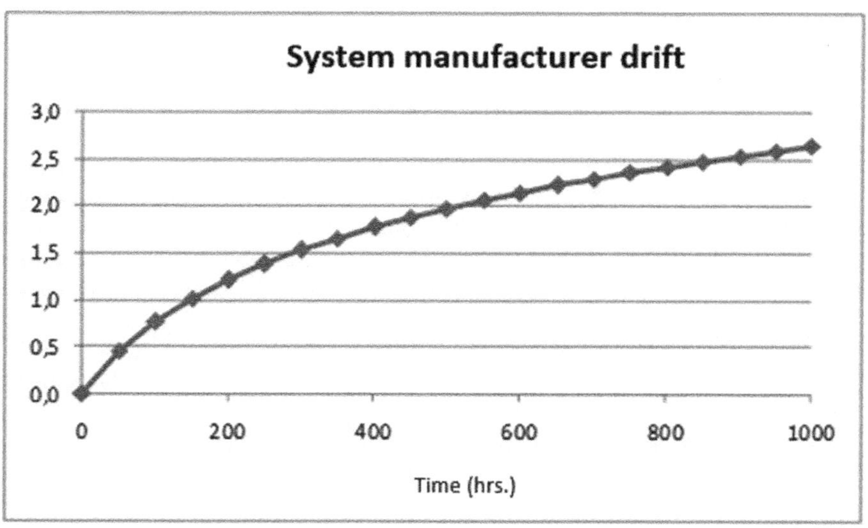

Figure 8.3. *Drift after 100 hours stabilization*

The specification requiring a maximal drift of 3 after 1,000 hours is then met since it is reached ~2.7 after 1,000 hours.

NOTE.–

To obtain this curve, the drift at 100 hours was subtracted from the drifts obtained for times exceeding 100 hours.

8.2. Stabilization

However, the problem is not that simple for three key reasons:

– The drift performance is achieved for a period of time equal to the product's duration of service, which is generally specified in years, or even in dozens of years. A long stabilization time would therefore be needed, which is incompatible with the industrial constraints of the equipment manufacturer.

In order to avoid this problem, with drift generally being greater the higher the temperature of the component, the test at the equipment manufacturer can be conducted at a temperature that exceeds that which the component is subjected to during operation. The virtual acceleration of time makes it possible to have an acceptable stabilization time. The disadvantage of this method is that an acceleration law must be estimated. This can only be done using tests at various temperatures (at least two).

– The parameters of the drift law (a and b in the previous example) can be very scattered from one component to another. This has two major consequences:

- the stabilization time cannot be unique and therefore common to all of the components;

- in certain cases, the stabilization time can be very long and therefore incompatible with the industrial constraints due to a too low drift speed. In this case, it is recommended to preserve the maximal authorized stabilization time and reject the component if this time is exceeded, in order to obtain the expected drift.

– The drift of the component cannot always be modeled with the drift law valid for the other components. In this case, the component is also rejected.

8.2.1. *Proposed principle*

The fact that the temperature used during the stabilization stage is higher than the temperature the product will be subjected to during use creates a problem; this is because the Arrhenius law is only valid for a constant temperature. To solve this issue, let us enter the world of Mr. Sedyakin. The principle he proposed for a survival function is also applicable to degradation. Assume the degradation can be written as follows:

$$D(t) = a.t^b \quad [8.2]$$

Assume the AFT (AF = acceleration factor) model can be used. The degradation can be written in the following forms:

$$D(t, X1) = a.t^b$$

$$D(t, X2) = a.(AF(X1, X2).t)^b \text{ with X2} > \text{X1}$$

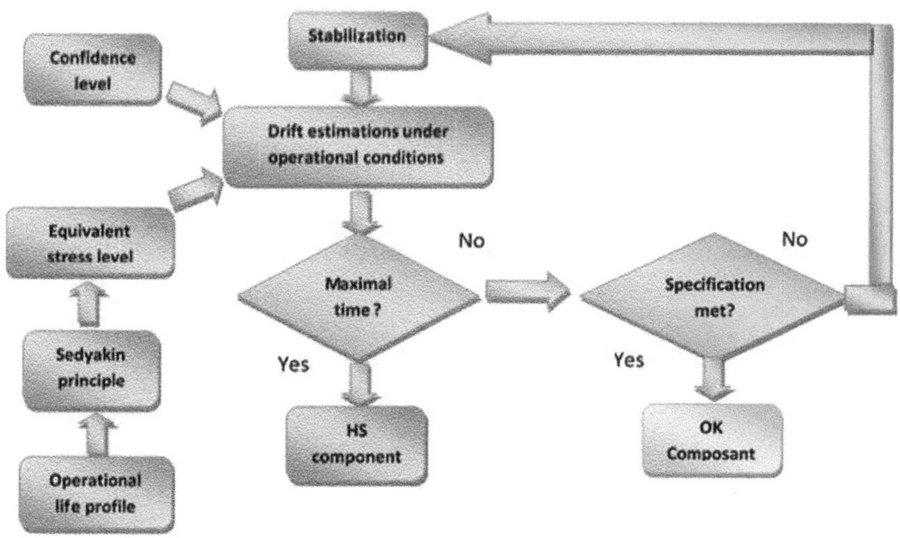

Figure 8.4. *Run-in principle. For a color version of this figure, see www.iste.co.uk/bayle/maturity1.zip*

Resuming the previous example, Figure 8.5 illustrates these equations.

With the formulated problem, the evolution of physical contribution illustrated in Figure 8.6 is applicable.

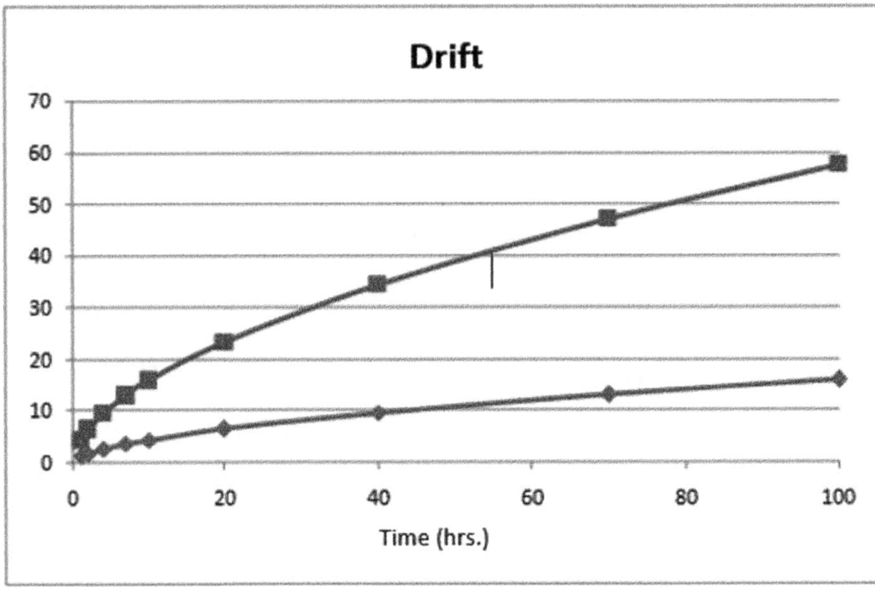

Figure 8.5. *Example of degradation with two different levels of physical contributions. For a color version of this figure, see www.iste.co.uk/bayle/maturity1.zip*

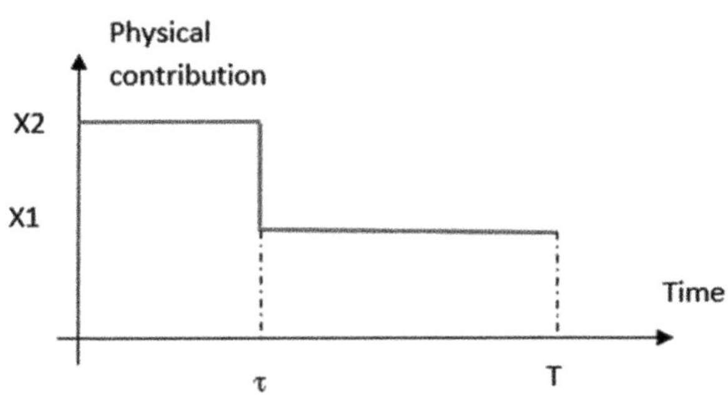

Figure 8.6. *Descending level of physical contribution. For a color version of this figure, see www.iste.co.uk/bayle/maturity1.zip*

The level of this physical contribution can be written as: $\mathrm{X} = \mathrm{X1}.\,\mathrm{I}_{t\leq\tau} + \mathrm{X2}.\,\mathrm{I}_{t>\tau}$. If the component is subjected to a physical contribution of level X2 during a time τ, then level X1, according to Sedyakin's principle, it can be written as:

$$D(\tau, X2) = D(t^*, X1) \qquad [8.3]$$

For this example, Sedyakin's principle can be illustrated by Figure 8.7.

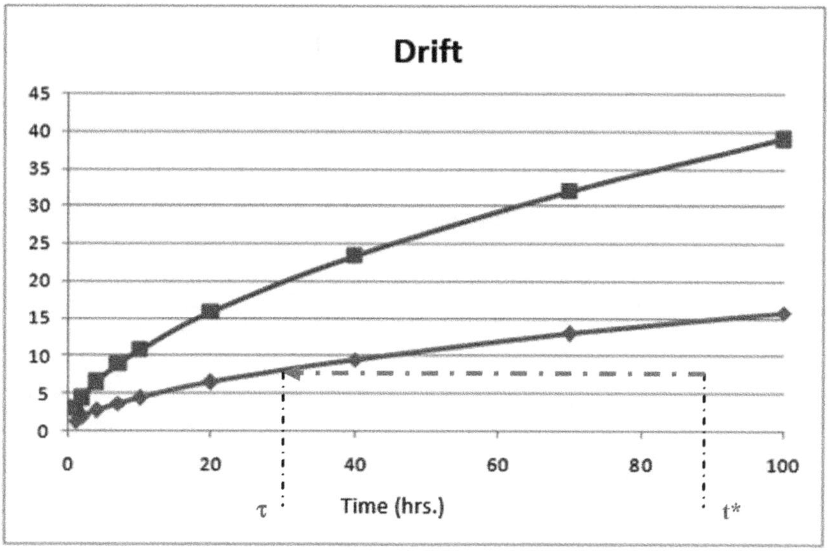

Figure 8.7. *Illustration of Sedyakin's principle for a descending level. For a color version of this figure, see www.iste.co.uk/bayle/maturity1.zip*

Equations [8.2] and [8.3] lead to:

$$D(\tau, X2) = D(AF(X1, X2).\tau, X1) = D(t^*, X1)$$

or still

$$t^* = \tau.AF(X1, X2) \quad [8.4]$$

8.2.2. *Drift acceleration law*

The Arrhenius law is generally used to model the effect of temperature (Arrhenius 1889). The insertion of this law into equation [8.1] leads to:

$$D(t, \theta) = \mathrm{a.ln(AF(\theta).t) + b} \quad [8.5]$$

where θ is the temperature of the component during the stabilization test.

AF is the acceleration factor between the accelerated conditions of the test and the reference conditions

NOTE.–

Time acceleration by a physical contribution (in this example, the temperature) can be modeled using the AFT (Acceleration Failure Time) model proposed in Bagdonavicius and Nikulin (2001).

According to the hypothesis of Arrhenius law, this takes the following form:

$$AF(\theta) = \exp\left(\frac{\mathrm{Ea}}{\mathrm{Kb}}.\left(\frac{1}{\theta_{\mathrm{ref}}} - \frac{1}{\theta}\right)\right) \quad [8.6]$$

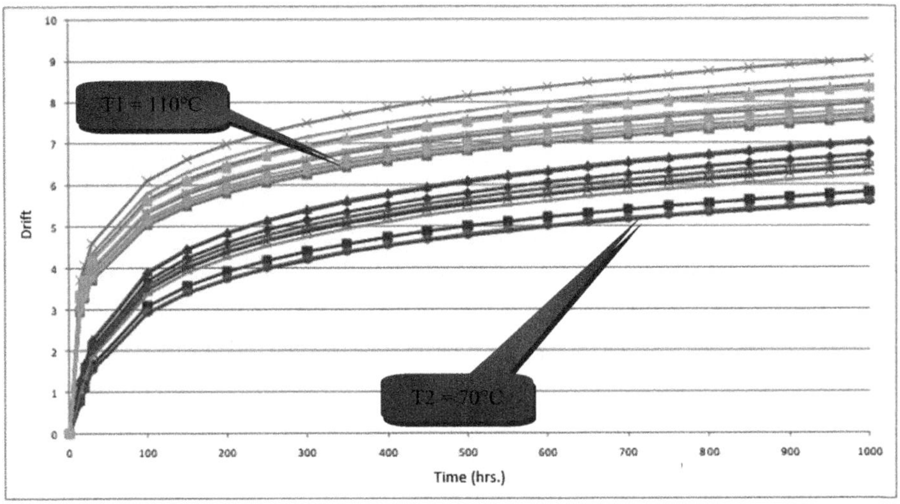

Figure 8.8. *Example of the effect of temperature for the drift of a performance. For a color version of this figure, see www.iste.co.uk/bayle/maturity1.zip*

Based on equations [8.2] and [8.3], the following can be written for the operational conditions:

$$D(t, \theta_{Test}) = \mathrm{a}.\ln\left(\exp\left(\frac{\mathrm{Ea}}{\mathrm{Kb}}.\left(\frac{1}{\theta_{\mathrm{opl}}} - \frac{1}{\theta_{\mathrm{Test}}}\right)\right).\mathrm{t}\right) + \mathrm{b} \quad [8.7]$$

To illustrate this, let us resume the example of a logarithmic drift with 10 components for two test temperatures θ1 = 110°C and θ2 = 70°C. Assume that the activation energy is Ea = 0.6 eV.

Figure 8.8 is obtained.

NOTE.–

A small random term was added in the expression of parameter "a" so that there are different trajectories between the components, thus closer to reality.

8.2.3. *Choice of the drift model*

Several models of drift are possible. The least squares regression line is the method that is typically used. It is important to note that it is the variable that depends linearly on the regression coefficients. For example, a power model is a linear model.

At first glance, several linear models of drift may be suitable. The coefficient of determination R^2 is often used to verify the quality of the regression. This is an error, for at least three reasons:

– The same coefficient R^2 for a case with five drift measures and another case with 50 measures does not necessarily provide the same information on the quality of the chosen model.

– What is the value of R^2 starting with which the contemplated model can be considered representative for the data?

– A model with many parameters has more chances to provide a more precise model in graphical terms.

In order to choose the best-fitted drift model, the use of the Akaike information criterion (AIC) (Akaike 1974) is proposed. The Akaike information criterion makes it possible to penalize models depending on the number of parameters in order to meet the parsimony criterion.

The chosen drift model is the one with the weakest Akaike information criterion. The overview diagram give in Figure 8.9 illustrates the proposed method.

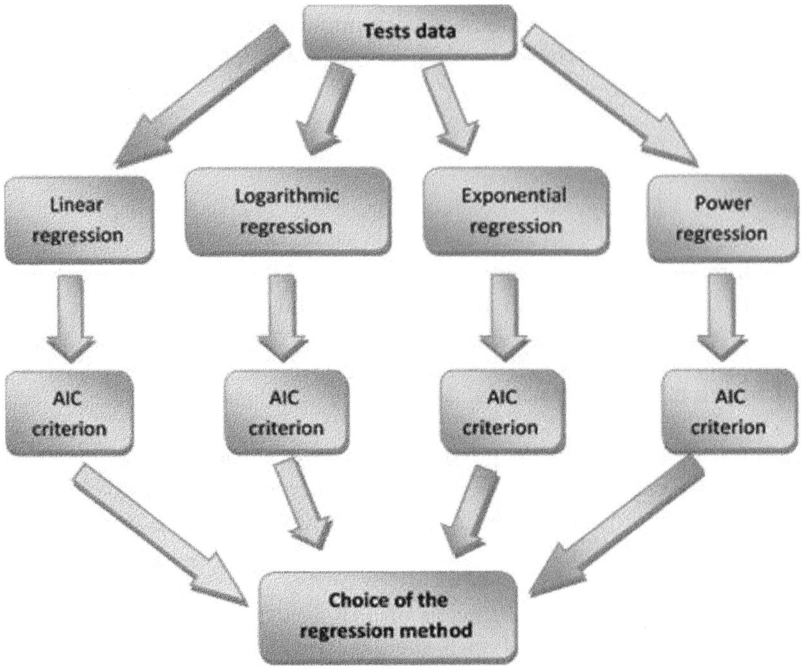

Figure 8.9. *Principle of the choice of the drift model*

The AIC criterion is defined as follows:

$$AIC = \mathrm{n}.\ln\left(\frac{\sum_{i=1}^{n}(D_i-\widehat{D_i})^2}{n}\right) + 2.(p+1) \qquad [8.8]$$

where n is the number of points of measurement; p is the number of parameters of the model; D_i is the drift measured at point "i" and $\widehat{D_i}$ is the drift estimated at point "i".

8.2.4. *Equivalent level of physical contribution*

To find the acceleration factor between the accelerated conditions of the test and the operational conditions, the operational life profile – to which the component is subjected – must be known. Most of the time, this is composed of the various operational stages for which the duration and the level of the various physical contributions are known. This division is in fact compulsory because the physical laws of failure (Arrhenius, Coffin–Manson, etc.) are only valid for constant levels. The division into various stages of the life profile makes it possible to apply these physical laws of failure.

It is important to be able to estimate the acceleration factor between the conditions of the stabilization test and the operational conditions for various physical contributions. Sedyakin's principle (SED 1966; Bayle 2019), more commonly referred to as "cumulative damage", is used for this purpose, in order to obtain an equivalent constant level of physical contribution in terms of reliability.

For the temperature case, which represents the majority of cases, this leads to:

$$\theta_{eq} = \frac{-\mathrm{Ea}/_{\mathrm{Kb}}}{\ln\left(\int_0^t \exp\left(-\frac{\mathrm{Ea}}{\mathrm{Kb}.\theta(u)}\right).du\right)} \quad [8.9]$$

When the thermal time constant of the product is low compared to the duration of the stages of life profile, the temperature can be considered constant during the stage. According to this hypothesis, the previous equation is written as:

$$\theta_{eq} = \frac{-\mathrm{Ea}/_{\mathrm{Kb}}}{\ln\left(\sum_{i=1}^{p} \frac{D_i}{\sum_{i=1}^{p} D_i}.\exp\left(-\frac{\mathrm{Ea}}{\mathrm{Kb}.\theta_i}\right)\right)} \quad [8.10]$$

where p is the number of stages of the considered life profile, D_i is the duration of stage "i" and θ_i is the temperature of stage "i".

Consider the example of a life profile involving 5 stages, as defined in Table 8.1.

Stage name	Duration	Temperature
Stage 1	820 hrs	20 °C
Stage 2	400 hrs	55 °C
Stage 3	1 250 hrs	43 °C
Stage 4	4 350 hrs	21 °C
Stage 5	3 300 hrs	38 °C
Total	10 120 hrs	

Table 8.1. *Example of calculation of equivalent temperature*

In this example, an equivalent temperature of ~ 34.3°C is obtained.

8.3. Expression of the corresponding degradation

Consequently, the degradation can be written as:

$$D(t,X) = \mathrm{D}(\mathrm{t},\mathrm{X1}).\mathrm{I}_{\mathrm{t}\leq\tau} + \mathrm{D}\left(\frac{\mathrm{t}-\tau+\tau.AF(X1,X2)}{AF(X1,X2)},\mathrm{X1}\right).\mathrm{I}_{\mathrm{t}>\tau} \quad [8.11]$$

PROOF.–

It can be written that:

$$D(t,X) = \mathrm{D}(\mathrm{t},\mathrm{X2}).\mathrm{I}_{\mathrm{t}\leq\tau} + \mathrm{D}(\mathrm{t}-\tau+\mathrm{t}^{*},\mathrm{X1}).\mathrm{I}_{\mathrm{t}>\tau}$$

Or still, according to equation [8.4]

$$D(t,X) = \mathrm{D}(\mathrm{t},\mathrm{X2}).\mathrm{I}_{\mathrm{t}\leq\tau} + \mathrm{D}(\mathrm{t}-\tau+\tau.AF(X1,X2),\mathrm{X1}).\mathrm{I}_{\mathrm{t}>\tau}$$

Or still, according to AFT model

$$D(t,X) = \mathrm{D}(\mathrm{t},\mathrm{X2}).\mathrm{I}_{\mathrm{t}\leq\tau} + \mathrm{D}\left(\frac{\mathrm{t}-\tau+\tau.AF(X1,X2)}{AF(X1,X2)},\mathrm{X2}\right).\mathrm{I}_{\mathrm{t}>\tau}$$

End

Considering a degradation with a power law of parameters a and b, the following relation results:

$$D(t,X) = \mathrm{a}.\mathrm{t}^{\mathrm{b}}.\mathrm{I}_{\mathrm{t}\leq\tau} + \mathrm{a}.\left(\frac{\mathrm{t}-\tau+\tau.AF(X1,X2)}{AF(X1,X2)}\right)^{\mathrm{b}}.\mathrm{I}_{\mathrm{t}>\tau} \quad [8.12]$$

Consider the following example:

– a = 2

– b = 0.56

– Tm = 10 000

– AF = 24.7

– τ = 299

Figure 8.10 is obtained.

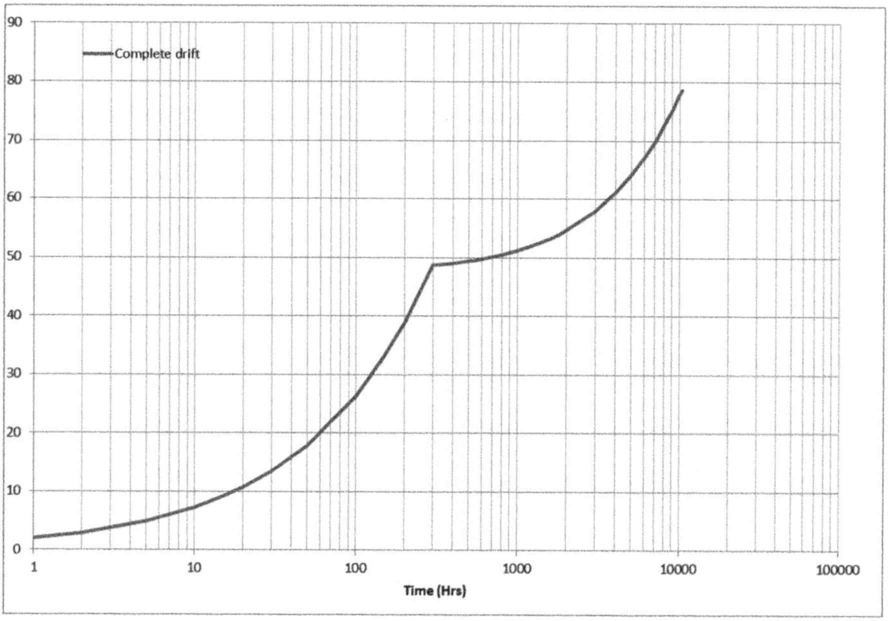

Figure 8.10. *Power law drift with temperature change. For a color version of this figure, see www.iste.co.uk/bayle/maturity1.zip*

8.4. Optimization of the stabilization time

Considering the previous example, there is a degradation, starting with time τ, of 48.653. If the specified objective is 30, this stabilization time is not appropriate. The value of the optimum stabilization time Dsta required for the operational drift to be lower than or equal to the specified maximal drift D_Obj should be determined.

Therefore:

$$\Delta D = \mathrm{D}\left(\frac{\mathrm{Tm}+AF(Xopl,Xacc).Dsta}{AF(Xopl,Xacc)},\mathrm{Xopl}\right) - \mathrm{D}(\mathrm{Dsta},\mathrm{Xacc}) \qquad [8.13]$$

where Tm is the duration for putting products into service; Xopl is the operational level of the physical contribution of interest and Xacc is the accelerated level of the physical contribution during the stabilization stage.

Equations [8.1] and [8.7] lead to:

$$\Delta D = \text{a.}\left(\frac{\text{Dsta.}AF(Xopl,Xacc)+Tm}{AF(Xopl,Xacc)}\right)^{\text{b}} - \text{a. Dsta}^{\text{b}} \quad [8.14]$$

Due to the concavity of degradation, $0 < b < 1$ and this equation has no analytical solutions. However, the stabilization time can be estimated using numerical methods. The objective is now to find the value of the stabilization time Dsta that minimizes the equation:

$$\left|\text{a.}\left(\frac{\text{Dsta.}AF(Xopl,Xacc)+Tm}{AF(Xopl,Xacc)}\right)^{\text{b}} - \text{a. Dsta}^{\text{b}} - \text{D\_Obj}\right| \quad [8.15]$$

Resuming the previous example leads to Figure 8.11.

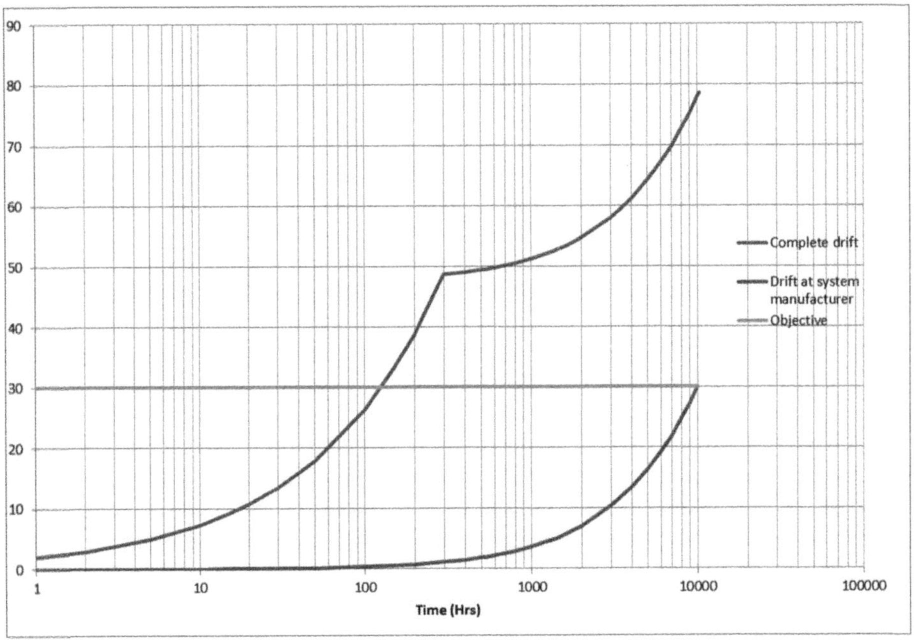

Figure 8.11. *Power law drift with change of temperature with optimum stabilization time. For a color version of this figure, see www.iste.co.uk/bayle/maturity1.zip*

The result is a minimum stabilization time of 299 hours to maintain the objective of maximal degradation of 20 under operational conditions.

8.5. Estimation of a prediction interval of the degradation

The linear regression theory is used here, and in particular the prediction interval. Indeed, in order to guarantee the preservation of frequency drift during operation with minimal risk, instead of the estimated value of drift, the higher bound of the prediction interval is taken into consideration.

I.e.:

– RSS (the residual sum of squares) measures the part of the Y variation unexplained by X, defined by:

$$RSS = \sum_{i=1}^{n} (Y_i - \hat{Y}_i)^2$$

– ESS (the explained sum of squares) measures the part of the Y variation explained by the model (by a variation of X), defined by:

$$ESS = \sum_{i=1}^{n} (\hat{Y}_i - \overline{Y})^2$$

The lower and upper bounds of the prediction interval (Rakotomalala 2018) at the confidence level CL are given by:

$$D_{\inf\_sup}(t) = D_{est} \pm St^{-1}\left(\frac{1+CL}{2}; n-2\right).\sqrt{\frac{RSS}{n-2}}.\sqrt{1+\frac{1}{n}+\frac{\left(\ln(t)-\ln(\overline{TTF})\right)^2}{ESS}} \qquad [8.16]$$

where n is the number of points of measurement; St^{-1} is the quantile of Student's law and t is the instant at which the value of the frequency drift is projected.

Resuming the previous example, Figure 8.12 is obtained.

It is clear that in order to estimate the minimal stabilization time, the upper bound of degradation must be taken into account.

8.5.1. *Principle of the stabilization method*

The principle of stabilization must meet several constraints, notably it must:

– guarantee a frequency drift below 1.5 Hz when the equipment is put into operation;

– have a stabilization duration that does not lead to excessively long test times.

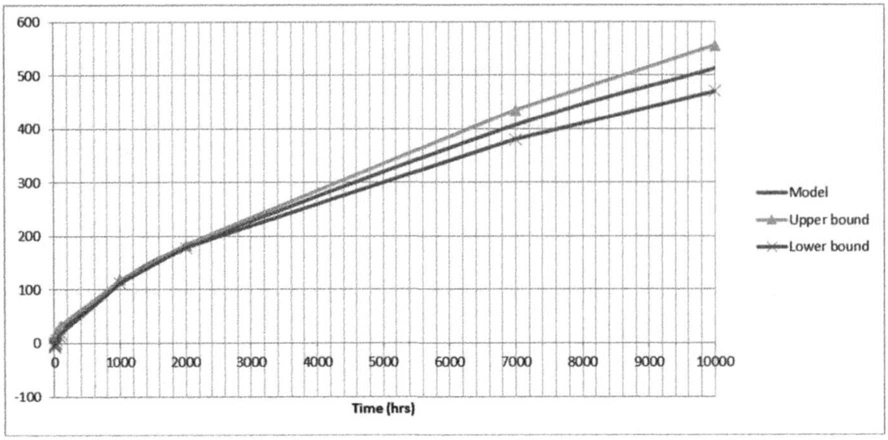

Figure 8.12. *Illustration of the prediction interval of a linear regression*

The principle relies on a methodology involving two distinct parts as follows, and illustrated by the following figure:

– A first series of tests allowing an estimation of the acceleration law of frequency drift. These tests were conducted on a sample of specific CDXO.

– For each CDXO delivered, a specific stabilization criterion had to be set.

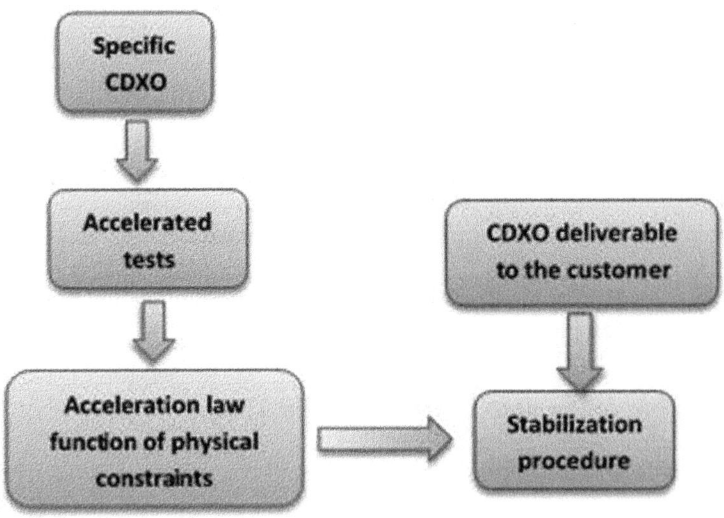

This part presents the modeling results obtained during the accelerated tests conducted on CDXO with the proposed method.

It is important to recall that the stabilization of CDXO to guarantee the preservation of the frequency drift specification involves two stages:

– A stage of estimation of the acceleration law in temperature and voltage conducted on a sample of dedicated CDXO.

– An optimization stage of the stabilization time to guarantee the preservation of the specification, which is specific to each CDXO delivered to the customer.

The parameters of the law of temperature drift are specific to each CDXO. Hence, in order to guarantee the preservation of the frequency drift with respect to the specification, it is necessary to optimize the stabilization time of each CDXO individually.

List of Notations

Ea	Activation energy
Kb	Boltzmann's constant
R(t)	Survival function or probability of proper functioning or reliability
Rocof	Rate of occurrence of mean number of failures per unit time
t	Calendar time
T	Random variable representing the failure instants
Tm	Mission duration
Vref	Reference voltage
Γ	Complete Gamma function
θ	Temperature (°C)
β	Shape factor of Weibull's law
η	Scale factor of Weibull's law
$\lambda(t)$	Instantaneous failure rate

ΔT	Temperature amplitude
$\bar{X}$	Arithmetic mean of variable X
$\hat{X}$	Estimator of variable X

List of Definitions

A

Accelerated test: a test during which the level of constraints or the frequency of their application exceeds the value corresponding to specified operational conditions, in order to reduce the time required to obtain a response to constraints.

Aggravated tests: an aggravated test is an accelerated test whose guiding principle is to bring an entity to a limit (structural or operating) behavior that renders it unfit for use, submitting it to progressively but relatively rapid increased stresses.

B

Batch: a group of homogeneous products manufactured and verified under the same conditions.

Bill of materials: the list of components achieving the functions of a product.

Burn-in: a manufacturing operation that involves submitting the concerned product, after fabrication, to physical contributions (climatic, electrical and mechanical) so that latent defects are rushed into obvious defects. The objective is to eliminate the products proven defective before delivery to the customer.

Burn-in effectiveness: the effectiveness of the process = percentage of latent defects revealed by burn-in. Economic effectiveness = cost effectiveness of the process, considering its benefits.

C

Capability: the capacity of a process to achieve characteristics that meet the requirements for these characteristics.

Carrier: a set of products integrated by each equipment manufacturer (e.g. airplane, car, train, rocket, satellite and smartphone).

Component: an entity (often an electronic device) that cannot be disassembled without being destroyed or its expected use being damaged (e.g. resistors, capacitors, diodes, integrated circuits and hybrid circuits).

Coverage rate: in this book, it designates the ratio between the equipment breakage rates detected by the tests and the overall failure rate of the products.

D

Defect: in this book, it designates non-conformity to a specification.

E

Entity: designates any element, component, device, sub-product, operating unit, material unit or product that can be considered individually.

Environmental robustness: the capacity of a product to maintain its operational performances in environments more severe than those specified, in order eliminate the manufacturing and usage variability.

Equipment: in this book, it designates a set of subassemblies that achieve a function or a group of specific functions. This equipment can be a subassembly of a product.

Equipment manufacturer: a manufacturer that designs and manufactures a product based on specifications. For example, in the automotive field, VALEO can be considered an equipment manufacturer.

F

Failure: the loss of the ability to function as required.

Failure mechanism: the physical process driving a failure.

Final user: a private person or professional using a product sold by a system manufacturer. For example, in the automotive field, any person that owns a Renault vehicle can be considered the final user.

L

Latent failure: a failure that is not observable.

Level of constraint: in this book, it designates the amplitude of the constraint.

Lifecycle: a series of identifiable manufacturing stages that an entity goes through, from its design to its disposal. For example, the classical lifecycle of a product comprises the following stages: feasibility and definition, design and development, manufacturing and putting into service, exploitation and maintenance, renovation during life time or service life extension and retirement or disposal.

M

Manufacturing: in this book, it designates an action or way to manufacture or produce products in an industrial manner.

Maturity: the state of a product that guarantees the expected reliability level, since it is put into operational service.

N

Non-conformity: the failure to meet a requirement.

O

Operating life: the time the product is used by the final customer.

Operational robustness: the capacity of a product to achieve operational performances that are above those required in the specified environments.

Operational test: designates the verification of critical parameters to guarantee the proper operation of a product.

P

Physical contributions: physical constraints refers to any physical quantity, normally applied to the product during its operational use or for tests, including the relevant aspects of design. Physical constraints are grouped into various families: thermal, electrical, thermal cycling, mechanical, humidity and chemical.

Product: the assembled entity whose reliability is studied. Generally, equipment.

Production: in this book, it designates the amount of products manufactured throughout a given period or during a manufacturing cycle.

Q

Quantile: the quantiles of a univariate, discrete or continuous variable are the values taken by the variable for probability values below the considered quantile, having a remarkable value, for example 3 tenths or 5 hundredths. They are also known as fractiles, a complete synonym depending on the context of usage, and they are the reciprocal values of the distribution function of the law of probability.

R

Requirement: a formulated need or expectation, usually implicit, or imposed.

Robustness: the capacity of a product to display performances that are above those required in specified environments.

Robustness margins: the difference of level of a physical contribution between the specification and the observed operational or environmental limit.

Run-in: process conducted with the sole intention to stabilize the parameters.

NOTE.– A run-in is accelerated conditioning, which is achieved by getting the entity to operate under its operating electrical charge at a high temperature, which is generally the maximal operating temperature compatible with the assigned thermal characteristics of the device.

S

Sampling: in this book, it designates the samples of production used to verify the conformity of specifications.

Service life: the time until a certain percentage of aging failures (0.1%, 1% or 10%) is observed.

System manufacturer: a manufacturer specifying a product to an equipment manufacturer and integrating it in its carrier. For example, in the automotive field, RENAULT can be considered a system manufacturer.

T

Test: in this book, it designates a method subjecting an element to studied conditions, in order to define if this element meets its specifications.

Test in combined environment: in this book, it designates an environmental test during which several constraints are applied simultaneously.

Testability: the measure of the easiness with which an entity can be tested.

List of Acronyms

AIC	Akaike information criteria
CTR	Current transfer ratio
DE	Dreaded event
EM	Electromigration
ESD	Electrostatic discharge
ESR	Equivalent serial resistor
ESS	Environmental stress screening
FIT	Failure in time
FMEA	Failure mode and effect analysis
FMECA	Failure modes, effects and criticality analysis
FPGA	Field-programmable gate array
HALT	Highly accelerated life testing
HASS	Highly accelerated stress screening
HCI	Hot carrier injection
HOV	High operating vibration

HV	High vibration
ISO	International standardization organization
LED	Light emitting diode
LTOL	Low temperature operational limit
MOS	Mechanical overstress
MTBF	Mean time between failures
MTTF	Mean time to failure
MTTR	Mean time to repair
NASA	National aeronautics and space administration
NBTI	Negative bias thermal instabilities
NFF	No fault found
PBTI	Positive bias thermal instabilities
PCB	Printed circuit board
POS	Proof of screen
ppm	Parts per million
QFN	Quad flat no-leads
QMS	Quality management system
RETEX	Return of experience
SMD	Surface mounting device
TBF	Time between failure
TC	Temperature cycling
TDDB	Time dependent dielectric breakdown

TS	Thermal shock
TTF	Time to failure
TTR	Time to repair
UTOL	Upper temperature operating limit

References

Akaike, H. (1974). A new look at the statistical model identification. *IEEE Transactions on Automatic Control*, 19(6), 716–723.

Arrhenius, S. (1889). Uber die Reaktionsgeschwindigkeit bei der Inversion von Rohrzucker durch Säuren. *Zeitschrift für Physikalische Chemie*, 4, 226.

AVAGO (2018). Calculate reliable LED lifetime performance in optocouplers. White paper, Broadcom.

Bagdonavicius, V. and Nikulin, M. (2001). *Accelerated Life Models: Modeling and Statistical Analysis*. Chapman and Hall/CRC Press, New York.

Barringer, P. and Barringer & Associates Inc. (2010). Database of typical Weibull shape and characteristic life parameters (wdbase) [No longer available online].

Bayle, F. (2019). *Reliability of Maintained Systems Subjected to Wear Failure Mechanisms: Theory and Applications*. ISTE Ltd, London and John Wiley & Sons, New York.

BNAE (2005). RG.AERO 000 29. Guide for defining and performing highly accurate tests. Guide, Bureau de normalisation de l'aéronautique et de l'espace, France.

Duncan, A.J. (1965). *Quality Control and Industrial Statistics*, 3rd edition. R.D. Irwin Inc, Homewood, IL.

ECSS (2013). ECSS-Q-ST-60C Rev. 2. Electrical, electronic and electromechanical (EEE) components. Standard, ECSS.

EDR (2000). EIAJ EDR – 4704. Guideline for accelerated endurance testing of semiconductor devices. Guideline, Technical Standardization Committee on Semiconductor Devices.

FIDES (2009). Méthodologie de fiabilité pour les systèmes électroniques – guide FIDES 2009, Édition A. UTE C80-811.

Firadec (2015). Datasheet.

Gaudoin O. (2010). ENSIMAG course, INPG Grenoble.

Gaudoin, O. and Ledoux, J. (2007). *Modélisation aléatoire en fiabilité des logiciels*. Hermès Science Lavoisier, Paris.

IEC (2005). IEC-TR-62240. Process management for avionics – Electronic components capability in operation. International Electrotechnical Commission.

IEC (2013). IEC 62506. Methods for product accelerated testing. International Electrotechnical Commission.

IEC (2014). IEC 60300-1. Dependability management. International Electrotechnical Commission.

IES (1984). Survey of screening effectiveness. Institute of Environmental Sciences.

IMdR (2000). RDF 2000 (UTEC 80810). Modèle universel pour le calcul de la fiabilité prévisionnelle des composants, cartes et équipements électroniques. Institut pour la Maîtrise des Risques.

ISACA Enterprise (2018). Capability Maturity Model Integration (CMMI). Version V2.0.

ISI (2015). NF EN ISO 9001:2015. Systèmes de management de la qualité – exigences.

ISO (2015). NF EN 9001:2015. Quality Management Systems – Requirements. International Organization for Standardization.

ISO (2018). IEC 31010:2019 – Risk management – Risk assessment techniques. International Organization for Standardization.

JEDEC (2006). Failure mechanisms and models for semiconductor devices. JEP122C. JEDEC Solid State Technology Association.

JEDEC (2014). *Temperature Cycling: JESD22-A104E*. JEDEC Solid State Technology Association.

Kececioglu, D. and Sun, F.-B. (1999). *Environmental Stress Screening: Its Quantification, Optimization and Management*. Prentice Hall, Upper Saddle River, NJ.

Kececioglu, D. and Sun, F.-B. (2003). *Environmental Stress Screening: Its Quantification, Optimization and Management.* DEStech Publications Inc, Lancaster, PA.

Lall, P., Pecht, M.G., Hakim, E.B. (1997). *Influence of Temperature on Microelectronics and System Reliability*. CRC Press, New York.

Landzberg, A. (1969). Reliability of controlled collapse interconnections. *Interconnections Reliability*, 13(3), 266–271.

McLean, H.W. (2009). *Halt, Hass and Hasa Explained: Accelerated Reliability Techniques*. ASQ Press, Milwaukee, WI.

Miner, M.A. (1945). Cumulative damage in fatigue. *Journal of Applied Mechanics*, 12, A159–A164.

Murata (2020). Datasheet.

NASA (2003). EEE-INST-002: Instructions for EEE Parts Selection, Screening, Qualification, and Derating. NASA.

NAVSEA (1991). NAVSEA-TE000-AB-GTP-010. Parts Derating Requirements and Application Manual for Navy Electronic Equipment. Naval Sea Systems Command.

Nichicon Corporation (N/A). General Description of Aluminium Electrolytic Capacitors [Online]. Available at: https://www.tti.com/content/dam/ttiinc/manufacturers/nichicon/PDF/aluminum.pdf.

NPFC (1988). MIL-HDBK-338B. Military Handbook – Electronic Reliability Design Handbook. NPFC.

NPFC (2013). MIL-STD-750F. Test Methods for Semiconductor Devices. NPFC.

Rakotomalala, R. (2018). Econométrie : la régression linéaire simple et multiple. Thesis, Université Lumière Lyon 2, Lyon.

Rausand, M. and Hoyland, A. (2004). *System Reliability Theory: Models, Statistical Methods, and Applications*. Wiley-Interscience, Hoboken, NJ.

RIAC (2015). HDBK-217PLUS. Reliability Prediction Models. Reliability Information Analysis Center.

Rigdon, S. and Basu, A.P. (2000). *Statistical Methods for the Reliability of Repairable Systems*. Wiley, New York.

RNC (2006). Apports des essais aggraves (dont les essais jumelés HALT/HASS) pour un système électrique et logiciel (SEL). Report, CNES, Paris.

StatXpert (2014). Rapport interne Thales StatXpert. Proposition de plan de contrôle par échantillonnage. Report, StatXpert.

Syed, A. and Kang, W. (2016). *Board Level Assembly and Reliability Considerations for QFN Type Packages*. Amkor Technology Inc., Chandler, AZ.

Toshiba. (2018). Reliability Information Handbook. Handbook, Toshiba Electronic Devices & Storage Corporation, Tokyo.

Tuinenga, P. (1995). *SPICE: A Guide to Circuit Simulation and Analysis Using PSpice*, 3rd edition. Prentice Hall, Englewood Cliffs, NJ.

Vishay (2018). Datasheet.

Index

M, N

O, P

Q, R

S, T

V, W

Other titles from

ISTE

in

Mechanical Engineering and Solid Mechanics

2021

CHALLAMEL Noël, KAPLUNOV Julius, TAKEWAKI Izuru
Modern Trends in Structural and Solid Mechanics 1: Static and Stability
Modern Trends in Structural and Solid Mechanics 2: Vibrations
Modern Trends in Structural and Solid Mechanics 3: Non-deterministic Mechanics

DAHOO Pierre Richard, POUGNET Philippe, EL HAMI Abdelkhalak
Applications and Metrology at Nanometer Scale 1: Smart Materials, Electromagnetic Waves and Uncertainties (Reliability of Multiphysical Systems Set – Volume 9)
Applications and Metrology at Nanometer Scale 2: Measurement Systems, Quantum Engineering and RBDO Method (Reliability of Multiphysical Systems Set – Volume 10)

LEDOUX Michel, EL HAMI Abdelkhalak
Heat Transfer 1: Conduction (Mathematical and Mechanical Engineering Set – Volume 9)
Heat Transfer 2: Radiative transfer (Mathematical and Mechanical Engineering Set – Volume 10)

PLANCHETTE Guy
Cindynics, The Science of Danger: A Wake-up Call (Reliability of Multiphysical Systems Set – Volume 11)

2020

SALENÇON Jean
Elastoplastic Modeling

2019

BAYLE Franck
Reliability of Maintained Systems Subjected to Wear Failure Mechanisms: Theory and Applications
(Reliability of Multiphysical Systems Set – Volume 8)

BEN KAHLA Rabeb, BARKAOUI Abdelwahed, MERZOUKI Tarek
Finite Element Method and Medical Imaging Techniques in Bone Biomechanics (Mathematical and Mechanical Engineering Set – Volume 8)

IONESCU Ioan R., QUEYREAU Sylvain, PICU Catalin R., SALMAN Oguz Umut
Mechanics and Physics of Solids at Micro- and Nano-Scales

LE VAN Anh, BOUZIDI Rabah
Lagrangian Mechanics: An Advanced Analytical Approach

MICHELITSCH Thomas, PÉREZ RIASCOS Alejandro, COLLET Bernard, NOWAKOWSKI Andrzej, NICOLLEAU Franck
Fractional Dynamics on Networks and Lattices

SALENÇON Jean
Viscoelastic Modeling for Structural Analysis

VÉNIZÉLOS Georges, EL HAMI Abdelkhalak
Movement Equations 5: Dynamics of a Set of Solids
(Non-deformable Solid Mechanics Set – Volume 5)

2018

BOREL Michel, VÉNIZÉLOS Georges
Movement Equations 4: Equilibriums and Small Movements
(Non-deformable Solid Mechanics Set – Volume 4)

FROSSARD Etienne
Granular Geomaterials Dissipative Mechanics: Theory and Applications in Civil Engineering

RADI Bouchaib, EL HAMI Abdelkhalak
Advanced Numerical Methods with Matlab® 1: Function Approximation and System Resolution
(Mathematical and Mechanical Engineering SET – Volume 6)
Advanced Numerical Methods with Matlab® 2: Resolution of Nonlinear, Differential and Partial Differential Equations
(Mathematical and Mechanical Engineering SET – Volume 7)

SALENÇON Jean
Virtual Work Approach to Mechanical Modeling

2017

BOREL Michel, VÉNIZÉLOS Georges
Movement Equations 2: Mathematical and Methodological Supplements
(Non-deformable Solid Mechanics Set – Volume 2)
Movement Equations 3: Dynamics and Fundamental Principle
(Non-deformable Solid Mechanics Set – Volume 3)

BOUVET Christophe
Mechanics of Aeronautical Solids, Materials and Structures
Mechanics of Aeronautical Composite Materials

BRANCHERIE Delphine, FEISSEL Pierre, BOUVIER Salima, IBRAHIMBEGOVIC Adnan
From Microstructure Investigations to Multiscale Modeling: Bridging the Gap

CHEBEL-MORELLO Brigitte, NICOD Jean-Marc, VARNIER Christophe
From Prognostics and Health Systems Management to Predictive

Maintenance 2: Knowledge, Traceability and Decision
(Reliability of Multiphysical Systems Set – Volume 7)

EL HAMI Abdelkhalak, RADI Bouchaib
Dynamics of Large Structures and Inverse Problems
(Mathematical and Mechanical Engineering Set – Volume 5)
Fluid-Structure Interactions and Uncertainties: Ansys and Fluent Tools
(Reliability of Multiphysical Systems Set – Volume 6)

KHARMANDA Ghias, EL HAMI Abdelkhalak
Biomechanics: Optimization, Uncertainties and Reliability
(Reliability of Multiphysical Systems Set – Volume 5)

LEDOUX Michel, EL HAMI Abdelkhalak
Compressible Flow Propulsion and Digital Approaches in Fluid Mechanics
(Mathematical and Mechanical Engineering Set – Volume 4)
Fluid Mechanics: Analytical Methods
(Mathematical and Mechanical Engineering Set – Volume 3)

MORI Yvon
Mechanical Vibrations: Applications to Equipment

2016

BOREL Michel, VÉNIZÉLOS Georges
Movement Equations 1: Location, Kinematics and Kinetics
(Non-deformable Solid Mechanics Set – Volume 1)

BOYARD Nicolas
Heat Transfer in Polymer Composite Materials

CARDON Alain, ITMI Mhamed
New Autonomous Systems
(Reliability of Multiphysical Systems Set – Volume 1)

DAHOO Pierre Richard, POUGNET Philippe, EL HAMI Abdelkhalak
Nanometer-scale Defect Detection Using Polarized Light
(Reliability of Multiphysical Systems Set – Volume 2)

DE SAXCÉ Géry, VALLÉE Claude
Galilean Mechanics and Thermodynamics of Continua

DORMIEUX Luc, KONDO Djimédo
Micromechanics of Fracture and Damage
(Micromechanics Set – Volume 1)

EL HAMI Abdelkhalak, RADI Bouchaib
Stochastic Dynamics of Structures
(Mathematical and Mechanical Engineering Set – Volume 2)

GOURIVEAU Rafael, MEDJAHER Kamal, ZERHOUNI Noureddine
From Prognostics and Health Systems Management to Predictive Maintenance 1: Monitoring and Prognostics
(Reliability of Multiphysical Systems Set – Volume 4)

KHARMANDA Ghias, EL HAMI Abdelkhalak
Reliability in Biomechanics
(Reliability of Multiphysical Systems Set –Volume 3)

MOLIMARD Jérôme
Experimental Mechanics of Solids and Structures

RADI Bouchaib, EL HAMI Abdelkhalak
Material Forming Processes: Simulation, Drawing, Hydroforming and Additive Manufacturing
(Mathematical and Mechanical Engineering Set – Volume 1)

2015

KARLIČIĆ Danilo, MURMU Tony, ADHIKARI Sondipon, MCCARTHY Michael
Non-local Structural Mechanics

SAB Karam, LEBÉE Arthur
Homogenization of Heterogeneous Thin and Thick Plates

2014

ATANACKOVIC M. Teodor, PILIPOVIC Stevan, STANKOVIC Bogoljub, ZORICA Dusan
Fractional Calculus with Applications in Mechanics: Vibrations and Diffusion Processes

Fractional Calculus with Applications in Mechanics: Wave Propagation, Impact and Variational Principles

CIBLAC Thierry, MOREL Jean-Claude
Sustainable Masonry: Stability and Behavior of Structures

ILANKO Sinniah, MONTERRUBIO Luis E., MOCHIDA Yusuke
The Rayleigh–Ritz Method for Structural Analysis

LALANNE Christian
Mechanical Vibration and Shock Analysis – 5-volume series – 3rd edition
Sinusoidal Vibration – Volume 1
Mechanical Shock – Volume 2
Random Vibration – Volume 3
Fatigue Damage – Volume 4
Specification Development – Volume 5

LEMAIRE Maurice
Uncertainty and Mechanics

2013

ADHIKARI Sondipon
Structural Dynamic Analysis with Generalized Damping Models: Analysis

ADHIKARI Sondipon
Structural Dynamic Analysis with Generalized Damping Models: Identification

BAILLY Patrice
Materials and Structures under Shock and Impact

BASTIEN Jérôme, BERNARDIN Frédéric, LAMARQUE Claude-Henri
Non-smooth Deterministic or Stochastic Discrete Dynamical Systems: Applications to Models with Friction or Impact

EL HAMI Abdelkhalak, RADI Bouchaib
Uncertainty and Optimization in Structural Mechanics

KIRILLOV Oleg N., PELINOVSKY Dmitry E.
Nonlinear Physical Systems: Spectral Analysis, Stability and Bifurcations

LUONGO Angelo, ZULLI Daniele
Mathematical Models of Beams and Cables

SALENÇON Jean
Yield Design

2012

DAVIM J. Paulo
Mechanical Engineering Education

DUPEUX Michel, BRACCINI Muriel
Mechanics of Solid Interfaces

ELISHAKOFF Isaac *et al.*
Carbon Nanotubes and Nanosensors: Vibration, Buckling and Ballistic Impact

GRÉDIAC Michel, HILD François
Full-Field Measurements and Identification in Solid Mechanics

GROUS Ammar
Fracture Mechanics – 3-volume series
Analysis of Reliability and Quality Control – Volume 1
Applied Reliability – Volume 2
Applied Quality Control – Volume 3

RECHO Naman
Fracture Mechanics and Crack Growth

2011

KRYSINSKI Tomasz, MALBURET François
Mechanical Instability

SOUSTELLE Michel
An Introduction to Chemical Kinetics

2010

BREITKOPF Piotr, FILOMENO COELHO Rajan
Multidisciplinary Design Optimization in Computational Mechanics

DAVIM J. Paulo
Biotribolgy

PAULTRE Patrick
Dynamics of Structures

SOUSTELLE Michel
Handbook of Heterogenous Kinetics

2009

BERLIOZ Alain, TROMPETTE Philippe
Solid Mechanics using the Finite Element Method

LEMAIRE Maurice
Structural Reliability

2007

GIRARD Alain, ROY Nicolas
Structural Dynamics in Industry

GUINEBRETIÈRE René
X-ray Diffraction by Polycrystalline Materials

KRYSINSKI Tomasz, MALBURET François
Mechanical Vibrations

KUNDU Tribikram
Advanced Ultrasonic Methods for Material and Structure Inspection

SIH George C. *et al.*
Particle and Continuum Aspects of Mesomechanics